Praise for *A Thousand Barrels a Second*

"Peter Tertzakian's analysis of world oil is a fascinating reminder that history often foretells the major turning points of the future."

—Gwyn Morgan,
President & Chief Executive Officer,
EnCana Corporation

"*A Thousand Barrels a Second* is an excellent book! In my more than 40 years in the industry I can't think of a publication that has so clearly discussed the global challenges of today's demands and tomorrow's requirements."

—Peter Gaffney,
Senior Partner,
Gaffney, Cline & Associates

"*A Thousand Barrels a Second* is a book that arrives just in time, providing a strategic assessment of our current situation and 10-year outlook. We can all benefit from its insights, and I recommend it to all global policy makers."

—U.S. Representative Charles F. Bass, (R-NH), member
House Energy and Commerce Committee

"In *A Thousand Barrels a Second*, Peter Tertzakian explains the truth behind the *real* energy crisis. The book is a fascinating portrayal of where the oil issue will take the world economy and American business in the next 15 years, and should be required reading for those of us in the real estate industry."

—Dave Liniger,
Chairman,
RE/MAX International

"*A Thousand Barrels a Second* provides unique historical context for the challenges we face in the energy arena. Peter Tertzakian draws fascinating parallels between past 'break points' in the energy industry and the current situation."

—Jon Erickson,
Managing Director,
Princeton University Investment

"Bravo to Peter Tertzakian for taking on a very complex and contentious issue—our society's near-addiction to oil—and doing a masterful job at describing the history, present circumstances and implications, and outlining rational strategies for the future."

—Gregory B. Jansen,
Managing Director,
Commonfund Capital, Inc.

"In *A Thousand Barrels a Second*, Peter Tertzakian lays out a vision of the future for producing as well as consuming nations, and issues a warning that while we will all survive, those who remain uninformed will pay a greater price."

—Hank Swartout,
Chairman,
Precision Drilling
Corporation

"Peter Tertzakian shines a very bright light on an enormously critical issue facing every business and every human being on the planet. I highly recommend this book."

—Ronald L. Nelson,
President and Chief Financial
Officer,
Cendant Corporation

"You can't lead today without a thorough understanding of how energy impacts people's lives. What makes *A Thousand Barrels a Second* great is that this complex subject is made perfectly clear."

—Phil Harkins,
CEO,
Linkage, Inc.

"Peter Tertzakian's perspective is both unconventional and uniquely qualified. As an experienced geophysicist, he understands the challenge of finding and developing new sources of crude oil and natural gas. As an economist and historian, he understands the context in which 'break points' occur. *A Thousand Barrels a Second* provides timely and valuable insight into the energy markets of today and tomorrow."

—Hal Kvisle,
Chief Executive Officer,
TransCanada Corporation.

A Thousand Barrels a Second

The Coming Oil Break Point and the Challenges Facing an Energy Dependent World

Peter Tertzakian

McGraw-Hill

New York / Chicago / San Francisco / Lisbon / London / Madrid / Mexico City
Milan / New Delhi / San Juan / Seoul / Singapore / Sydney / Toronto

1 2 3 4 5 6 7 8 9 0   DOC/DOC   0 9 8 7

ISBN-13: 978-0-07-149260-7
ISBN-10:   0-07-149260-7

This is a paperback edition of ISBN 0-07-146874-9

McGraw-Hill books are available at special quantity discounts to use as premiums and sales promotions, or for use in corporate training programs. For more information, please write to the Director of Special Sales, McGraw-Hill Professional, Two Penn Plaza, New York, NY 10121-2298. Or contact your local bookstore.

This book is printed on acid-free paper.

CONTENTS

ACKNOWLEDGMENTS

This book culminates over two decades worth of readings, analyses, thoughts, and experiences. Assembling ideas, researching historical anecdotes, generating themes, threads, and historical metaphors, giving speeches, trampling through the bush searching for oil and gas, analyzing reams of numbers and charts, and even examining antiques and artifacts filled my head with immeasurable content. Of course none of this knowledge accumulation and synthesis was in isolation. Over the years I have collaborated and debated with countless peers, colleagues, clients, and anyone else who has cared to listen, challenge, and form my ideas. I thank all those who have influenced my thinking along the way.

"I couldn't have done it myself," is an understatement. Phil Harkins' patient ears and no-nonsense advice made me serious about tackling a project of this magnitude. Coaching me, encouraging me to think big, and introducing me to the right people were all crucial to this project.

Early on Phil Harkins introduced me to Keith Hollihan, whose creative mind, skilled wordsmithing, and enthusiasm for the subject matter helped me turn fragments of complicated subject matter into easily digestible reading material. Sounding board, editor, researcher, writer, and friend; Keith's contributions to this book were second to none.

Behind the scenes my partners at ARC Financial have been enormously supportive at every step. Many took the time to read drafts of the book, suggest changes, and point out errors. Active discussion at our weekly research meetings, where we review and discuss energy trends, helped keep my facts straight and my thoughts unbiased. Thanks to all at this exceptionally talented firm. And special thanks to my long-time, close colleague Kara Baynton who during seven years has contributed immensely to this book through research, pragmatic advice, and patience. Among her many contributions, Kara read and commented on the manuscript an unreasonable number of times.

Kind endorsements from Gwyn Morgan, Peter Gaffney, Dave Liniger, Greg Jansen, Ron Nelson, Hank Swartout, Charles Bass, Hal Kvisle, Jon Ericksen, and Phil Harkins were humbling.

Many thanks to B.G. Dilworth of the B.G. Dilworth Agency, who offered no-nonsense advice on the structure of the book, secured the publisher, and meticulously edited the text. Finally, gratitude goes to Jeanne Glasser at McGraw-Hill, who believed in the project when energy issues were not making daily, front-page news. She and her colleagues at McGraw-Hill helped put 20 years of my abstractions onto the bookshelf.

Finally, the members of my family who inspire me: My wife Janet, and our two sons Alexander and André. Patiently, they have allowed me to pursue this project on top of a full-time job. This book is dedicated to them.

THE COMING OIL BREAK POINT

"America is addicted to oil," President George W. Bush declared in his January 2006 State of the Union address. At the time, it seemed a stunning admission for an oil-friendly president to make. A year later, it's clear that what the president said was an understatement. The U.S. oil addiction is symbolized by the country's SUV-jammed freeways, far-flung suburbs, vulnerability to foreign sources of supply, and growing consumption year over year. But this is not just a U.S. story. On the global front, nearly every one of the 192 nations on this planet is also an oil addict, and newly industrializing countries in Asia and the Middle East are growing their habits as quickly as "substance abusers" can. All told, the world now consumes oil at the staggering rate of *a thousand barrels a second*. As a milestone, this is a nice even number, but think about it in immediate terms. By the time you finish reading this paragraph, over 200 thousand gallons of gasoline will have been burned. Next year our rate of consumption will be greater.

Our growing addiction to oil is not easily sustainable, if in fact it is sustainable at all, and this means that big changes in the world of energy are coming at you faster than you think.

As you read this book, we are on the cusp of a tipping point—what I call a *break point*—that will change the way governments, corporations, and individuals exploit and consume primary energy resources, especially crude oil. Over the course of the next 5 to 10 years, increasingly volatile energy prices are going to affect how you live and what you drive, not to mention the economy, the environment, and the complex geopolitical chess match that is now being played out for the world's precious energy resources.

The most tangible example of how this energy break point will affect our lives occurred in 2005, in the aftermath of Hurricanes Katrina and Rita. When Gulf Coast drilling platforms, refineries, and pipelines stopped feeding the United States, the sudden jump in costs, the desperate calls for action, and the anxious feeling of economic and even political insecurity were reminiscent of the oil shocks of the 1970s and early 1980s. Like any addict, we became extremely aware of the vulnerability of the supplies of energy that light our homes, turn our wheels, and power our cities and industries.

The fragility in our energy lifeline that Katrina and Rita exposed highlights the increasingly untenable balance between the way we are supplied with oil and the way we consume its marvelous products: day-to-day necessities like gasoline, heating oil, and jet fuel. Future potential calamities—natural or political—will continue to exacerbate the pressure points of an intertwined global problem. "America is addicted to oil" was only the first half of President Bush's statement. He went on to say that our oil "is often supplied from unstable parts of the world." This caveat, along with increasing concerns about global warming, shows how complicated resolving our addiction has become.

The challenges that can potentially aggravate our oil vulnerabilities are many. At different moments, we will blame geopolitical flashpoints, natural disasters, unsustainable consumption, tightening supplies, or the slow boil of global warming. In fact, the surface causes are largely irrelevant, since they are all symptomatic of the overall oil break point. Our primary focus should be on our response to this break point. Serious structural and lifestyle changes will be necessary. Beginning now, and over the course of the next 5 to 10 years, we will be forced to come to grips with our oil addiction and rally to a new balance in our energy use.

Making a convincing case for that statement, and for the timeliness of this book, might have been a challenge two or three years ago. After all, few people in society, business, or government worried about long-term trends in the energy industry, as prices were low, global warming was a non-issue, and energy security wasn't perceived to be threatened. But events over the course of the past 24 months have revealed the warning signs of change.

The ongoing chaos in Iraq and the 2006 war in Lebanon highlighted the fragile and fractious temperament of politics in the Middle East, a region that holds the largest bounty of oil on the planet. China and India's voracious, growing appetite for energy to feed their addiction now has those consumer giants competing with the United States for the increasingly difficult-to-find oil that remains. In fact, "multipolar" competition for oil, which was a major driver of geopolitical tensions in the 1920s, has returned in the past couple of years and is slowly gnawing away at the western world's thinning sense of energy security.

Our vulnerability to political turmoil is not limited to the Middle East. Nigeria, the world's sixth largest exporter of oil, is increasingly plagued by civil strife and armed rebellion, mainly targeting that country's offshore oil platforms. Russia, the second largest producer of oil in the world after Saudi Arabia, continues its not-so-subtle campaign to nationalize its oil industry. Not lost on Russia, or on any other major producer that wants to make a mark on the world stage, is the important notion that political power and control of oil go hand in hand in an energy-addicted world.

As if there weren't enough issues encumbering the world's oil arteries, two new themes have gained greater prominence recently: nuclear proliferation and climate change. Iran's determination to acquire atomic capabilities adds yet another layer of tension in the Middle East, a complication that promises to fester for a long time and to have dangerous implications for the world oil supply. On the environmental front, global warming and the demand for cleaner energy highlight our vulnerability to supply cost, another aspect of the overall pressure leading toward the break point.

Indeed, in many industrialized countries, calls by politicians and the public to do something about global warming have grown louder and louder. Burning fossil fuels—namely, coal, oil, and natural gas—is conjectured to be a root cause of climate change. But what alternatives do we have? As this book will explain, nothing in the energy world comes for free, and no new sources of energy have the capability to fundamentally replace the massive industrial complex of fossil fuels. If we as a society are serious about mitigating climate change, we must reconcile ourselves to the fact that this will be either a costly endeavor or one that requires substantial changes in lifestyle.

Price volatility is an indicator of the coming break point, one that's easy to spot when costs are rising. People have a tendency to relax and think that all is well when prices fall, even though a downward drop reinforces the existence of volatility. In mid-2006, the price of crude oil reached new highs approaching $80 per barrel, hurting corporate profits and creating uncertainty about the future for many industries.

By the end of that summer, gasoline prices started falling rapidly. Did that mean that our problems were over? Hardly. The high hurricane activity that had been anticipated for 2006 did not occur, the war in Lebanon ended, and backroom diplomacy was being given another chance to resolve the Iranian nuclear standoff. To the oil markets, which thrive on the specter of natural disaster and the threat of conflict, it seemed as though world peace and harmony had broken out. In addition, the U.S. economy began to slow down, and the frenzied rate of global economic expansion seemed to have finally peaked. For the first time in four years, the tension in the world's oil supply chains appeared to be loosening instead of tightening. As a result, gasoline prices fell back to the low two-dollar range. But this did not mean that our circumstances had improved fundamentally, or that the problem of our addiction had been resolved.

Nor is the news about other primary energy commodities any better, as the prices of natural gas, coal, corn, and uranium have at least doubled since 2002. The reason for this wide-sweeping change is simple: worldwide, the demand for energy is growing at a never-before-seen pace, just as supplies of inexpensive, light sweet crude are finally tightening and getting more difficult to find. The impact is only beginning to be felt. As the pressure builds, we will soon wake up to the realization that the age of cheap, clean, easy-to-obtain energy is rapidly coming to an end.

Regardless of how our temporary circumstances fluctuate, core challenges that include unbridled consumption growth, increasing supply vulnerability, rising supply costs, and a constrained set of alternatives are resolutely leading us to a break point. Metaphorically, the pressure may have eased temporarily, but the pot is still boiling and is ready to repressurize on short notice, an event that will occur with great likelihood over the next decade.

Because of the urgency of these issues and the breadth of their impact, I felt compelled to write a book that would explain the dynamics of this world-changing event. The chapters that follow represent my highly researched and balanced assessment of our energy situation. While I am not needlessly alarmist about the extent of the changes that will be visited on us, I am realistic about the uncertainty and volatility that we will experience in the years to come. Although the stakes have never been greater, the history of energy shows that a time of crisis is always followed by a defining break point, after which government policies, and social and technological forces, begin to rebalance the structure of the world's vast energy complex. This dynamic is already underway.

Break points are crucial junctures marked by dramatic changes in the way energy is used. During the break point (and the 10- to 20-year rebalancing phase that follows), nations struggle for answers, consumers suffer and complain, the economy adapts, and science surges with innovation and discovery. In the era that emerges, lifestyles change, businesses are born, and fortunes are made.

This book is about understanding solutions and seizing opportunities as the looming oil break point approaches, even as it deciphers the myths and realities of today's headlines about the energy industry. As an earth scientist who once explored for oil, a history buff and entrepreneur who appreciates the changes that technological innovation have brought to our society, and a chief economist and investment strategist who tracks traditional and alternative energy issues, my job is to look into the future and provide advice to those making multimillion-dollar decisions. The questions that business leaders, politicians, and concerned citizens have for me are simple but profound: How high will the price of oil go? Why are these changes happening? Are we running out of resources? What will happen to the world's economies? Where are the solutions going to come from? How can we take full advantage of the opportunities?

In providing answers, I examine many dynamic variables, including the economy, the weather, technological advances, environmental issues, social factors, policy strategies, and geopolitics. Most of these factors have long been taken for granted because energy has been available to us without undue pain or worry for the last 25 years. But even now, a new lexicon of issues has become fodder for popular debate: Is China's growing thirst for energy sustainable? Have we entered a new multipolar world in which energy is the primary source of global tension? Are biofuels like ethanol a panacea for growing U.S. gasoline consumption? Will nuclear power and coal save the day—again? Will you really be driving a fuel cell vehicle in the next decade, and will it even matter? Which government policies work and which do not? What sort of global landscape will emerge from the turmoil? How can individuals and businesses navigate the next volatile decade? Where will the real—as opposed to the wished for—opportunities be found?

The issues are confusing even to the experts. With my team, I sift every day through a constant barrage of news releases, numbers, and charts to turn the chatter and white noise into substantive ideas, forecasts, and recommendations. This book is about today and the future. However, the more I look for long-term clarity, the more I am drawn to the past. As a society, we have come to expect that rapid technological change will always meet our needs and solve our problems. And while the energy industry is as high tech as any in the world, it remains rooted in decisions made generations ago. Only by examining history is it possible to fully understand our current situation and find solutions for the future.

In this way, I will take you through a journey of growing understanding about energy. As you read this book, it is my hope that you will gain insight into

- The way historical choices have created entrenched pathways and difficult-to-displace standards that severely limit the options available to us today
- The geopolitical currents that have inspired a global scramble to stake out energy claims with an intensity that we have not seen since just after World War I; and the fundamental issues concerning our most valuable fuel, crude oil, that are launching us into a new era of volatility and a subsequent quest for balance
- How environmental and political concerns color our choices, and why new age technologies will not provide the magic bullet to solve our near-term difficulties

In the end, my message is a cautiously positive one: there are energy options available to us, many of which will be surprising and unexpected to most readers. Understanding these possibilities will inspire confidence and optimism in our ability to navigate the future.

Will the fuel cell become the steam engine of tomorrow? What will the next Edison be discovering in his or her laboratory? Where will the General Electric or Standard Oil of tomorrow emerge? Will the struggle for oil between the United States and China define the next generation of geopolitics the way the struggle between the United States and Britain defined the early twentieth century?

Someday, historians will mark the first two decades of this century as the dawn of a new energy era.

CHAPTER 1

LIGHTING THE LAST WHALE LAMP

We're not running out of oil. There is plenty of oil left in the ground to last us many decades, if not longer. We are, however, running short of cheap oil, especially the desirable grade of oil that flows easily and is devoid of sulfur, otherwise known as "light sweet crude." Our reliance on that cheap oil runs deeper and is more entrenched than most of us are aware, and because its supply is getting tight at a time when global demand is accelerating, a great change is underway that will put pressure on our lifestyles and our world. This book is about those pressures and why they will be so difficult to resolve. But it's also about the light at the end of the tunnel. Understanding the history of how we arrived at this point will help us to know what's coming in the next couple of decades, and it is through such knowledge that each of us, as individuals, business leaders, and citizens, will make smarter decisions. It may even turn a few of us into the Edisons and Rockefellers of a new energy era.

Every time we flick on a light switch, turn up the heat, or start up our car, a vast and complex energy supply chain kicks into gear. To fuel and power our lifestyles, the world in 2005 draws from these supply chains to consume 85 million barrels of oil, 240 billion cubic feet of natural gas, 14 million tons of coal, and 500,000 pounds of uranium *every single day*. Light sweet crude is only one part of that energy mix, but despite a lot of effort and wishful thinking to remove it from the equation, it remains the most crucial element. Our thirst for it is insatiable.

Historically, light sweet crude has been found in large fields. It's cheap because it's relatively easy to extract, transport, and refine. But gushers like the famous Texas Spindletop, discovered in 1901, reportedly spewing oil at a rate of 75,000 barrels per day, are simply not turning up very often any more. The odd one that does is usually offshore in deep ocean waters, or in some politically charged region like the Middle East.

Over the course of the last 145 years, and certainly in the last 30, geologists and geophysicists have mapped the planet extensively. We've used all sorts of high-tech remote sensing techniques, from satellite telemetry to high-resolution seismic signals to do so. I took part in this search as a high-tech foot soldier for Chevron Corporation in the early 1980s. We'd set up camp in remote, uninhabited areas of Canada's north, fighting off mosquitoes and black flies so relentless that we'd still hear their buzzing in our ears long after we'd laid down to sleep. Working long days, we'd survey the territory, bulldoze the trees in cut lines, and explode dynamite in carefully drilled holes to take acoustic soundings of the geological formations below the surface. That data was processed using supercomputers back at the home office, where other geologists, geophysicists, and engineers interpreted the subsurface maps to make million dollar decisions about where to drill.

Since the early 1990s, more and more of this imaging has been done using advanced 3D seismic technology, creating a virtual reality picture of what lies below the surface. Today, many historically prolific oil-producing areas like Texas, Oklahoma, and western Canada, have been "imaged" in substantial detail. Even the deep oceans have been mapped this way. Talk to any petroleum geologist or geophysicist today and you will hear the same thing. Nearly all the really big "elephant" oil fields, the ones that contain billions of barrels of reserves, have been identified.

So what's left to find? Aside from a handful of oil-rich regions, today's oil fields are increasingly smaller in size. A new oil field containing a few hundred million barrels of reserves is big news. At the current rate of global consumption, such fields would be drained in days if we could turn on a spigot. Moreover, many of these new reserves are located in geographically and politically inhospitable regions, generally the last places on earth to be mapped in great detail. If I thought my experiences 25 years ago in the wilds of northern Canada were tough, believe me, I would hate to be part of an oil exploration team now. Chances are, I'd be stationed in deep offshore waters, a remote desert, or in some uninviting jungle filled with rebel soldiers toting machine guns.

Another consideration is that not all oil is created equal. When newspapers and newscasts quote the price of oil, they are referring to the highly desirable light sweet grades like West Texas Intermediate or North Sea Brent, which are easily refined into gasoline. The infrastructure of pipelines and refineries around the world have historically been built with this grade of crude in mind. But today, when experts are talking about new oil fields or increasing production output levels, they are also referring to lesser quality, heavier, and more tar-laden grades of oil.

Given the technical difficulties and the risks involved in extracting such oil, the price has to be pretty high to make it worth exploring for, then bringing it out of the ground, and building pipelines and facilities to move it to market. At $20 per barrel—the inflation-adjusted price that we became accustomed to over the last thirty years—there are few places left on the planet where the economic incentives justify independent oil companies to find and drill any new wells. And that's only the supply side of the story. With global demand for oil rising every year, global production declining in the absence of massive investment, and with over one billion new consumers in China awakening with their own powerful thirst, the world is going to need every extra barrel of oil the industry can find. Sometime in 2006, mankind's thirst for oil will have crossed the milestone rate of 86 million barrels[1] per day, which translates into a staggering one thousand barrels a second! Picture an Olympic-sized swimming pool full of oil: we would drain it in about 15 seconds. In one day, we empty close to 5500 such swimming pools.

Considering the steadily growing demand, the resulting logic is grim: higher oil prices are needed to provide the incentive for exploration; over time most of the new oil fields are getting smaller, more costly to evaluate, and more risky to tap into; therefore prices need to go higher and higher to keep up the incentive to explore. Oil at $20 per barrel is history, at least until major changes reduce the uncertainty, pressure, and volatility that we are now only beginning to experience. Reasonable experts—including myself—believe that oil prices are going to become increasingly volatile over the next few years. Seasonal spikes of $100 per barrel or more could easily be the new reality that consumers may have to bear until changes are made.

Nevertheless, the daily news about oil is arbitrary, contradictory, and confusing. We're told many different things, often based on misconceptions or half-truths. For instance, we've all heard that OPEC[2] can produce more oil and bring down the price, or that drilling in ANWR, the Alaskan nature preserve, will alleviate U.S. dependency on Middle East oil. Other pundits claim that a new Manhattan Project[3] can wean us off oil altogether, while many consumers have come to believe that hybrid cars and fuel cells are the answer or that conserving electricity will have a direct impact on oil consumption. None of these magic bullets are practical now, or will make a difference any time soon.

In fact, our problems aren't going to go away for a decade or more. North American addiction to cheap energy is too strong, and the technological standards of the last century too entrenched, for any new or different approach to be easily or painlessly (let alone quickly) adopted. Moreover, because of its rapidly growing demand for imported oil, the United States is becoming increasingly exposed to global risk. Right now, the only thing anyone cares about is the rising price of energy; but soon we'll be worried about potential changes to our lifestyles, the trade-off between cheap energy and clean energy, the necessity of building new refineries and power plants in our own backyards, and even the impact on national security. Our birthright of abundant, reliable energy is coming to an end.

Why is this happening? How will we find a way forward to a cheaper, cleaner, more secure energy future? The answers are complex, but they're also fascinating. Throughout history, because of our evolving energy needs, we've gone through cyclical periods of protracted demand increases, volatile tension and pressure in our supply chains, followed by a break point that ultimately provokes great innovation and change in the structure of the world's energy sources. We call this "the energy cycle." During high-pressure eras such as today in which a break point is imminent, we'll go to any lengths to secure the energy we need—scavenging, hoarding, and even engaging in war for resources that spike in price. The balance returns only when consumption patterns change, and new energy resources or processes are discovered and restructured into the economy. Getting back to a point of balance is never easy, but it can be made less painful if we understand the dynamics and evolution of the energy cycle.

Pressure, Break Point, Rebalance

Most of us are savvy to the idea that there are booms and busts in the business cycle. We've seen ups and downs in the overall economy, and in narrow sectors like real estate, jobs, stocks, bonds, and even commodities like oil and gold. The idea that our fortunes rise and fall with an almost seasonal rhythm is ingrained in us from biblical times, and modern economists have put forth models to explain and track the regularity of this pattern. Some of those models are exceedingly complex and data intense, others more simple.

But what about the energy cycle? In fact, there are many small cycles within the overall energy market. For decades, as you may even be aware, every time the price of an energy commodity like coal, oil, or natural gas has gone up, broad market mechanisms have brought the price down again. In simple terms, as prices rise, producers rush more supply to the market, such as when OPEC announces an increase in its daily production of crude oil to meet demand. At the same time, during price hikes, people and industries have a tendency to pare back their consumption. In conjunction, these two responses allow prices to go back down again. Conversely, when prices are too low, people and industries have a tendency to use energy wastefully. As a simple example, consider how gas-guzzling vehicles like SUVs and Hummers emerged as popular driving choices in the late 1990s when a gallon of gas cost less than a gallon of milk; in contrast, after the energy crisis of the 1970s, we had been conditioned by high prices to buy small, fuel-efficient cars like the Pinto. In addition, during low-price eras, industries do not have incentive to focus on efficiency or conservation, and energy companies have no interest to invest in more production. As a result, supply becomes pinched, prices eventually go back up, and the wheel turns again.

That's a very basic interpretation, and we know the dynamics are more complicated than that, but it helps to imagine this cycle at play throughout the decades. Figure 1.1 illustrates a long-term model of how our energy systems evolve over time—from wood stoves to nuclear power plants, to whatever may come next. Although it looks innocuous enough on the page, let me explain the dynamics to show where we are now, and why tumult and uncertainty are going to be the norm for the next several years.

Start at the top of Figure 1.1, Growth and Dependency. Every economy, from the agrarian age to the modern era, uses more energy as it grows. Whether the energy that the economy relies on derives from wood, coal, or crude oil, those primary resources are exploited as the economy expands, energy consumption increases, and dependencies form. Indeed, whenever a new energy source or carrier takes root in a society, a frenzy of new products and services proliferate to take advantage of the opportunities. As an obvious example, consider how the development of electricity led to countless electronic devices from the toaster to the CAT scan.

Eventually, the primary energy resource becomes scarce, and pressure begins to build. A variety of forces can contribute to the intensity of the pressure, including environmental concerns, geopolitical competition, social trends, policy decisions, and business behaviors. Today, for example, concerns over the environment

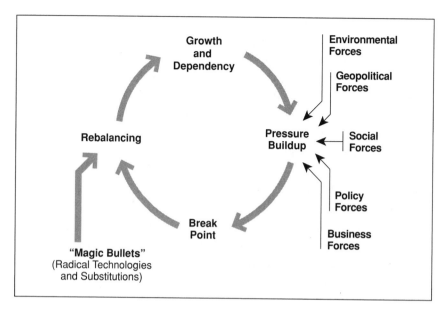

Figure 1.1   Energy Evolution Cycle

impose barriers on tapping into coal reserves or drilling in nature preserves, creating greater reliance on existing resources. Meanwhile, geopolitical competition between China and the West has created a global scavenger hunt for new energy reserves. Consumer behavior, such as the trend toward large, gas-guzzling automobiles, has put additional pressure on energy supplies. Pro-growth government policies rather than pro-conservation ones have contributed to the mounting crisis. And businesses in the private sector are making their own market-based decisions, adding to the strain on current energy capacity levels.

Sometimes these forces rebalance themselves relatively painlessly, but as global oil consumption tops one thousand barrels per second, it is clear that we are now approaching a dramatic break point in the energy cycle whose consequences will reach into every home. Even a relatively manageable break point period like the oil shocks of the 1970s reverberated worldwide for almost 15 years until conservation policies and the introduction of new energy sources rebalanced the supply and demand equation. In comparison, today's predicament has the potential to be longer, more confusing, and unmanageable because there are no radical technologies or simple fuel substitutions available to solve our current issues.

Much of this book is devoted to understanding the factors that are leading us to the break point; the rest describes the transition we will take to rebalance our energy needs and position ourselves for the next phase of growth. Radical technologies, a national rallying cry, aggressive tax and incentive policies, an authoritarian crackdown on consumer behavior—these are the kinds of approaches that have catalyzed a major rebalancing before. No matter what approach is taken, history has shown us that even a decade is a fast leap in the energy industry. In the meantime, we will all suffer through the uncertainty and difficulty of the transition until a new energy balance is found.

Lighting the World

We've been through such transitions before. The story of energy is an often dramatic and turbulent tale of world events and social evolution driven by the economics of supply and demand, the build-up of pressure on valued resources, and the "magic bullets" of ingenious innovation. Today, the world is lit and powered by a mix of fuels, including coal, uranium, crude oil, natural gas, and renewables like wind and solar power. But just 150 years ago whale oil was the world's primary illuminating fuel. If you think that our search for crude oil has been extensive and intense in recent decades, imagine a time when men chased whales across the oceans to meet the world's growing energy thirst. Indeed, from its rise in the mid-1700s to its peak in the mid-1800s and through its sudden and rapid decline in 1870, the whale hunt was more than a mere fishery; it was an ever more desperate search for the oil that lit up our world.

That story begins simply enough. For hundreds of years, Native Americans had caught whales off the coasts of Long Island and Cape Cod. They boiled the blubber on shore and used the oil as a preservative for hides and in their maize and beans. The early European settlers followed suit in the mid-1600s after realizing that whale oil made an excellent illuminant, far superior to the reed lamps or tallow candles they had long relied on for light. They also discovered that the oil was a great lubricant for their tools and farm equipment.

A small-scale whale fishery grew. When whales were spotted off the coast, small boats with six-man crews were launched to give chase. If the crew was lucky, the men would be able to harpoon the whale, lash it to the sides of the small boat, and drag it to shore at low tide. Huge "try pots" for boiling the whale blubber and rendering it into oil would be waiting for them on the beach, the fire under the pots lighting the way home in the dark.

At first, any kind of whale would do. Blackfish and humpbacks occasionally drifted too close to shore and were captured. Sperm whales, although rare to beach, were highly valued because their oil burned with a soft, clean light and a particularly fragrant smell. But right whales were initially the most prized catch for a simple reason: The baleen, or whalebone, found in the upper jaws could be fashioned into the rigid but flexible hoops needed in women's corsets which were popular at the time.

Eventually women's fashion took a backseat to energy, and demand for the sperm whale came to dominate the whaling industry. The sperm whale rush began in 1751 in Newport, Rhode Island, the day a merchant named Jacob Rodriguez Rivera wandered onto the docks to purchase a waxy substance known as spermaceti found only in the head of the sperm whale. Evidently, Rivera had entrepreneurial leanings since three years after emigrating from Spain and settling in Newport, he decided to go into the candle-making business. He began using spermaceti as his raw material, an idea that would revolutionize the candle industry and launch many a ship after the prized sperm whale.

The importance and utility of candles cannot be overstated. For thousands of years, during the course of the long agrarian era, candles were the means by which humans lit up the world. That made tallow, the grease or fat of animals used in making candles, one of our most important early fuels. Today, we may not think of a candle as being a way of storing and using energy, but it is, after all, just a solid fuel surrounding a wick. By the 1700s, candles came in many different forms and levels of quality. The simplest were made by dipping a rush or reed into kitchen grease. Slightly more expensive domestic candles were made from bullock or beef tallow, while cheaper ones were made from pig's fat, which created a great deal of black smoke and a foul smell. Sheep or mutton tallow was valued for its solidity and gloss, but because it was so costly it was often mixed with bullock tallow to reach a compromise between price and quality.

In our own era, gasoline is such a ubiquitous fuel that governments get a great deal of revenue from taxing it. Similarly, in their heyday, candles were such a valuable commodity that a British Act of Parliament applied a tax in 1709 and banned candle-making at home without a license to control production. Manufacturing became increasingly standardized as demand rose. The invention of the "dipping frame" made it possible to make many candles at once, while higher-quality candles were made in moulds that gave them a finished look. Still, even the best tallow candles "guttered" grease along their sides, producing a great deal of smoke and stink. Beeswax candles burned brighter and with a more pleasant smell, but making them was labor intensive since the wax could not be pressed in moulds but had to be ladled onto a wick and rolled by hand. Most people simply couldn't afford them.

And yet, even in the few short decades left before the industrial revolution began, the worldwide growth of commerce, trade and wealth was creating a powerful demand for more and better light. Rivera filled this need when he came up with his technique for producing candles from spermaceti. Before Rivera's discovery, the waxy spermaceti was usually mixed indiscriminately with the whale oil and blubber and boiled down. Now, Rivera, and the candle makers who followed him, wanted only the spermaceti. In an increasingly industrialized process, they learned to press spermaceti in burlap sacks and mix it with potash to remove the oil, creating a hard, white substance with a flaky, crystalline texture.

The resulting candles were expensive, but they were far superior to any other candles of the day. In fact, the bright white flame of the spermaceti candle would become the standard by which we would measure the quality and intensity of light well into the age of the electric light bulb.

With demand for spermaceti candles growing rapidly, a number of candlemakers went into business in New England, close to the whale supply: In Newport, candlemaking was dominated by merchants like Rivera and his son-in-law, Aaron Lopez. In Providence, a man named Benjamin Crabb set up shop with the support of a Quaker merchant named Obadiah Brown. In Massachusetts, Josiah Quincy, a Boston merchant whose capital came from the spoils of a captured Spanish ship, expanded his chocolate mill and glass factory in Braintree to include a candleworks, bringing his brothers-in-law, Joseph Palmer and Richard Cranch, in to run it.

Because of the superior quality of spermaceti, the sperm whale became the new prize catch of the whale fishery. Since sperm whales were much larger than right whales and lived in deeper waters, the ship-building industry responded with ever-larger and sturdier vessels. Still, the rarity of a catch meant that the price for spermaceti was volatile, depending on supply and making for an uncertain business. In the competition for this precious resource, the candleworks of Massachusetts and Rhode Island found themselves at odds with each other, as well as with the merchants who sold sperm whale oil, and the whalers and whale ship investors who wanted to sell their catch at the highest price.

To better manage the situation, the eight dominant candle manufacturers in New England, including Jacob Rivera's firm, joined to form the world's first energy cartel called the United Company of Spermaceti Candlers. The members of the cartel decided to share information about the marketplace. What's more, they agreed to fix the ceiling price for spermaceti and fix the floor price for the sale of candles. If high prices for spermaceti threatened their livelihood, the candle makers would pool their resources and go into the whaling business themselves. They also agreed to dissuade new candle-making firms from entering the business. Eventually, they even designed to treat the entire amount of spermaceti taken in by the American whaling fishery as common stock, which they would purchase through designated agents and divide among themselves in agreed-upon proportions.

The cartel didn't work. The market for whale oil and spermaceti was too dynamic with too many ambitious and competitive players to sustain. The whale oil merchants were trying their best as middlemen to control oil distribution too, and the whalers played oil merchants and candle makers against each other. One dominant oil merchant even tried to become vertically integrated as whaler, oil seller, and candle maker—an advance in management innovation that foreshadowed the rise of the twentieth-century oil conglomerate.

These contentious attempts by the candle makers and oil merchants of New England to control their industry resemble the conflicts between producers, suppliers, and consumers in our energy industry today, 200 years later. While consumers and governments often complain about the high oil prices realized by OPEC and the independent oil companies, they fail to recognize that producers and suppliers need to secure a price that supports the future cost of doing business. During eras of intensifying pressure, this conflict is one indicator of an approaching break point, the point when we realize that the ways and means by which we harness energy must undergo change. In fact, it was only a matter of time before something had to give in the whaling industry.

Chasing the Whale

After being temporarily interrupted by the American Revolution, the sperm whale hunt soon resumed its momentum, but with a new emphasis on whale oil over the waxy spermaceti.

For candle makers like Jacob Rivera, whale oil was just a byproduct of spermaceti refinement. But as a fuel, sperm whale oil actually had greater utility. It was flammable, but not so flammable as to be explosive. Like the spermaceti candle, it burned with a light that was bright in intensity but also soft and pleasing to the eye, producing a nice, sweet smell like "early grass butter in April,"[4] as Herman Melville put it in *Moby Dick*. And whale oil was even easier to transport than candles and also adaptable to a number of different devices, from house lamps to street lamps and even the lamps of lighthouses. It could also be used in textile mills for cleansing wool and lubricating machinery, and in the construction industry as a base for paint. Petroleum has a similar broad utility today, which is one reason why it is so difficult to displace no matter how high the price goes.

Spermaceti candles had arisen as an industry in 1750 because of an innovation in processing. In that same year, the production of sperm whale oil received its own boost from another innovation. For the first time, the try works—those large pots on the beach or near the docks in which whale blubber was boiled and rendered to oil—were installed on the whaling ship itself. This was a critical leap in technology. As whale stocks near the Northeast coast depleted, whale ships needed to travel for longer and longer voyages to find their catch, but the blubber rotted if it was not processed quickly, and the resulting oil was of a degraded quality and not very marketable. With the addition of the try works, whale ships no longer needed to return to shore after capturing a whale but could process it and store the oil in barrels with only a brief interruption of the hunt. Self-sufficient for longer periods of time, and with increasingly specialized work to do on board, whale ships began to resemble offshore drilling and production platforms.

One of the most famous ships in the history of the American whale fishery was the *Charles W. Morgan*. Named after its primary owner, the Morgan was a 351-ton ship launched in 1841 from New Bedford, Massachusetts, a town that rivaled Nantucket as a center for whaling. *The Morgan* took its first sperm whale four months after launch just off Cape Horn, Africa. A year later, it returned to port, carrying $54,686 worth of cargo at then market prices. After six months in port, it set sail again, and over the next 80 years of operation, *The Morgan* would record 37 voyages.

It survives today docked in Mystic Seaport, Connecticut, as the last remaining vessel of the American whaling fleet.

Charles W. Morgan himself was 45 years old at the time his ship first set sail, and he had already been in the whaling business for almost 20 years. As an investor, he had good reason to build and launch *The Morgan*, even though he already managed nine other whaling ships. Two years before, the price of sperm whale oil had reached its highest peak since the War of 1812. With scarcity of supply an ongoing certainty, the high cost of oil warranted further investment. Other ship investors were compelled by the same logic. *The Morgan* was one of 75 ships launched in New Bedford in 1841. Within a year, the American fleet would number an astounding 678 ships. This dramatic expansion of whaling capacity, inspired by normal market forces, helped temporarily bring down the pressure on the whale oil supply.

Just as oil drilling crews must go to incredible extremes today, so too whalemen experienced great hardship as the whale oil industry approached its break point. The voyages lasted up to four years, and promised boredom, backbreaking toil, brief moments of terrible danger, and exposure to disease, harsh weather, harsh treatment, poor food, crowded quarters, and bad smells. Seasickness was a common affliction. Bad health got worse without good medical care. Violent storms and violent encounters with whales threatened men's lives. In between those extremes, the quiet times on board could be rich in serenity or mired in anxiety, depending on the ship's recent fortunes. A ship that went a long time without spotting a whale was unlucky, and the men could hardly be blamed for feeling empty in the pockets.

Those men came from all over, a wide collection of races and nationalities. Some men left ship or died during the middle of a voyage, while other men were hired on at distant ports. In general, whalemen were looked down upon in society as dirty laborers who were eager for easy wealth over honest work. But they were drawn to the life for complicated reasons. Some saw the possibility of a quick fortune, however remote, as a way to establish themselves back on land with a farm and a wife. Others were turning away from an old life to start fresh, driven by a longing to see the world and be tested by all that nature could throw at them. The romance of the sea was strong. Poetic, philosophical, and religious thoughts were quick to come to those who contemplated the vastness of the rolling waves and the dark skies that flickered with far-off storms.

When a sperm whale was spotted, a cry went out, and the whaleship raced to get alongside it. With the whale in proximity, the whaleboats were lowered and the crews rowed fiercely to maneuver into place and launch a good strike. Harpoons with lines were thrown and sunk into the whale's side. As the whale lashed and twisted, the whaleboats struggled to keep the lines untangled and not lose the catch. When the whale finally exhausted itself, the whaleboats waited for the ship to pull closer so that the whale could be secured to the ship's side, and the hard work of slaughtering could begin.

The men stood on a wooden platform or cutting stage projected over the sperm whale's body. They cut the head off first and hoisted it to the ship, then began to strip the carcass of its blubber. The sailors who worked on the head used long ladles to retrieve the liquid spermaceti, and even climbed inside to get as much out as possible. The try works was fired up, and the spermaceti and blubber were boiled in the pots. It was hot, smoky, greasy work that could go on all night, the ship's deck lighting up the darkness with bright flames. The oil, while still warm, was barreled, and the casks were stored in the hold, the hatches battened. The job finished, the ship was cleaned and scrubbed and all the tools and ropes stored away, until no sign of the slaughter, grease, and smoke remained. In fact, the sperm whale oil had a restorative property to it that left the deck wood gleaming.

Not all whale oil was graded the same. In the marketplace, supply helped dictate price, but quality was also a crucial factor. When a ship docked, the buying agents tested each barrel of oil before deciding on a price to offer. The best-quality oil was the liquid purified from the spermaceti, which burned cleanest with a nice smell and commanded the highest price. A midgrade was assigned to oil rendered from the blubber of the sperm whale because it was not as pure or clean burning, but could still be sold as an illuminant. The lowest grade was given to oil that was rendered from right whale blubber, which produced more smoke when burned and was better suited as a lubricant for machinery. A low grade was also assigned to oil from sperm whale blubber that had gone rotten before it could be rendered in the try works.

This issue of quality was crucial for the fuel consumer in the early days of petroleum, too. Notably, when John D. Rockefeller began to sell kerosene as an illuminant, he named his company Standard Oil as a way of assuring customers that the quality of his product met a certain standard. Similarly, crude oil is traded in different grades today. As the supply of light sweet crude has tightened around the world, oil companies have been forced to invest extensively in the production and refinement of heavier, lower-grade oil like that extracted from the tar sands of Alberta, Canada.

By the time of Herman Melville's own experiences as a whaleman, sperm whale oil and spermaceti had become the great fuel of its day. As Melville wrote with flourish, the sperm whale was responsible for "almost all the tapers, lamps and candles that burn round the globe"[5] while kings and queens were coronated with the stuff, and the street lamps of London, the world's brightest city, were lit by it. With such demand, it's no surprise that people worked hard at refining the technology associated with the fuel. Soon, a new technological innovation created a means of burning whale oil more safely.

Sandwich, Massachusetts, was the oldest town on Cape Cod, and the center of whale lamp production. For oil lamp manufacturers, the ongoing technological concern was how to apply an effective stopper or threaded cap, which would allow oil to burn but not spill. The problem was not insignificant. If a lit lamp filled with whale oil tipped over, the flames spread so quickly that a house or factory could soon be engulfed. The whale oil itself wasn't the problem, but the makers of alcohol-based illuminants like camphene preyed on these fears with claims that their products were safer. This was false advertising at its worst, because such substances were actually far more explosive than whale oil.

The solution, as simple as it was ingenious, came in 1844 when a double-tube threaded cap was patented by Deming Jarves of the Jarves' Sandwich Glass Manufactory. Two tubes provided a second chamber to catch the whale oil and prevent it from spilling should the lamp be tipped. Now, whale oil was not only bright, clean, and nice smelling, but it could be burned safely, too.

Ironically, at the peak of its worldwide demand, the days of whale oil as a premium fuel were nearly over. Somehow, in 1847, Charles Morgan must have sensed the break point coming. At the very least, he recognized a bubble in the marketplace, and a good time to sell off his own financial interests in his namesake ship.

Morgan owned whale ships and candleworks and sold oil to lighthouses, but he was primarily an investor, the kind of man who was beginning to grow very wealthy in America. With the money he made whaling, he had already invested in mines, factories, mills, railroads, and finance companies of the sort that were emerging with the rise of the industrial age. Not unlike a savvy follower of tech stocks during the peak of the high-tech bubble in the 1990s, Morgan saw the writing on the wall for the whale industry. The price of whale oil was extremely high, while the whales themselves were becoming more scarce. Financing another voyage seemed like an increasingly risky prospect. What's more, the California Gold Rush had created great demand for ships, putting a premium on their price. After several attempts, Morgan finally managed to secure a deal and unloaded his ship to a man who wanted to get into the whale fishery.

He did so just in time. In 1849, Abraham Gesner, a Canadian geologist, distilled bituminous tar to produce coal oil. Gesner called the substance kerosene as a way of easing its adoption to those already familiar with the suffix in camphene. Kerosene was a wonderful new illuminant, as clean burning as whale oil and much cheaper, though not as nice smelling. Eight years later, with the invention of the kerosene burner by Michael Dietz in 1857, kerosene became the most sought-after illuminant on the market. Aside from the wealthy, most consumers could not afford sperm whale oil anymore. Factories and homes that relied on whale oil were reverting to tallow candles or unstable fuels like camphene to stay lit. No wonder kerosene, which cost only pennies a pint, was so readily embraced. Indeed, changing from a whale oil lamp to a kerosene-burning lamp was simple. One merely had to unscrew the double-tube cap of the whale lamp and replace it with a kerosene burner.

Rebalancing to this new fuel was all the more remarkable because so little trouble was required to adapt our hardware infrastructure. Today, as we look for radical new technologies or fuels to solve our problems, we need to consider how readily or successfully they can be adopted. The success of that switch-over depends on such things as price, quality, and how easy it is for consumers to make the change. When Thomas Edison introduced his electric light bulb in the 1880s, for instance, he made sure that the base of the lamp was designed so it fit the coal gas burners already in place in homes and businesses. In this way, Edison ensured that the shift, or rebalancing, to an alternative technology with an entirely new infrastructure of energy supply was not only cheap for consumers in homes and businesses, but as easy as screwing in a light bulb.

Sperm whale oil had one remaining advantage over kerosene: we knew where to find it in sufficient quantities. But the whale fishery was brought to a stunningly quick end when it was discovered that kerosene could be extracted and refined from rock oil, a greasy bitumen or "mineral rubber" liquid that oozed from the ground in the Oil Creek area around Titusville, Pennsylvania, and further north in Canada in the Eniskillen gum beds of Lambton County, Ontario.

Soon, entrepreneurs and industrialists turned their minds to figuring out how they could gather it up in greater quantities to meet the world's demand. Charles Tripp founded the International Mining and Manufacturing Company in 1851, digging a mostly uneconomic well in Ontario's Eniskillen gum beds. It was left to James Millar Williams, a shrewd carriage maker from Hamilton, Ontario, to buy Tripp's holdings and drill the first successful oil well in North America in 1858, though it did not initially go deep enough to yield much oil. Ontario would soon become an early hub of North American oil production, but prolific flows of subsurface oil would be exploited near Titusville first. In 1859, "Colonel" E. L. Drake, hired by a group of investors as their lead man in the field, took the idea of the derrick used to bore for salt, and adopted it to drill for oil on an artificial island right on Oil Creek. He used a steam engine to power the bit and bore through the earth. No one knew if Colonel Drake's plan would work, but when he struck oil 70 feet or so beneath the surface, he was able to pump it up by hand until it overflowed the nearby barrels and tubs. Like the plume of a whale's spout, this flow of oil would inspire another hectic race, drawing men to a new gold rush centered in Pennsylvania. In fact, some of the whalemen who had hunted the sperm whale would find themselves working the derricks as early wildcatters. The oil may have changed, but they were still chasing the whale.

The Last Whale Lamp

The new fuel, kerosene, was cheap enough to be afforded by nearly everyone. The conversion from whale oil to this new petroleum-based fuel marked the beginning of a sense that cheap, clean energy is our birthright, something we can take for granted. Kerosene's use as an illuminant was actually short-lived due to the introduction of the electric light bulb. But because of other timely innovations, namely diesel engines for ships and gasoline engines for automobiles, the crude oil from which kerosene was extracted soon became the most sought-after substance on the planet. In the 140 years since the U.S. whale fishery began its demise, our thirst for crude oil has gotten ever stronger, even as we supplemented our energy needs with coal, natural gas, hydroelectricity, and nuclear power.

Since the dawn of the modern era, our hunt for fuel has been a frantic one, catalyzed by the insatiable needs of our energy-hungry world. If the psychologist Abraham Maslow could append on his theory of the Hierarchy of Needs, he would do well to include energy, along with such basics as food, water, and shelter, as a primary need that must be fulfilled before other, higher needs get our attention. Energy is the underlying force that has shaped our history and built our modern world, even as it makes our society work. To see to that need, we have chased whales across the ocean, drilled into the depths of the earth, fought wars and fought them again, changed the course of rivers, and split the atom. With amazing ingenuity, we have created the means of converting fuel into the energy we need to light and power our lives.

Today, as the global supply of light sweet crude tightens and the demand for it continues to grow, our world is under great pressure, not unlike it was in the last days of the sperm whale fishery. Change is coming, as no less a figure than Alan Greenspan, Chairman of the U.S. Federal Reserve Board, noted in 2004, when he said: "If history is a guide, oil will eventually be overtaken by less-costly alternatives well before conventional reserves run out. Indeed, oil displaced coal despite still vast untapped reserves of coal, and coal displaced wood without denuding the forest lands. Innovation is already altering the power source of motor vehicles, and much research is directed at reducing gasoline requirements" He went on to say, "Nonetheless, it will take time. We, and the rest of the world, doubtless will have to live with the uncertainties of the oil markets for some time to come."[6]

Although Greenspan omitted whale oil from his historical overview, he got the trajectory right, but his sanguine comment about the time and ease it will take to navigate uncertainty misses the proverbial elephant in the room. We live in an age when technological change is rapid and seems to touch every aspect of our lives. But in the energy industry, the pace of radical change is slowing, not speeding up. Since the industrial age, we have made only five large-scale "alternative" substitutions—from wood to coal to whale oil to crude oil to natural gas to nuclear power. The only radical innovation in the entire twentieth century was nuclear power, a source of energy that most people, especially Americans, prefer not to rely on. At present there is nothing radically new on the horizon, no magic bullet that can topple the compelling utility of a primary energy source like oil. Any truly novel solutions we do come up with will take decades to implement. Nothing is as simple as screwing in a new burner any more. Radically new energy substitutes to help rebalance will not come at us from outside the evolutionary cycle[7], as say rock oil did in 1859 and nuclear power did in 1957. Indeed, to mitigate our current oil dependency, we will have to find rebalancing solutions that come from within the confines of known energy supply chains, and from within the ruts of the established evolutionary cycle.

And yet, I have no doubt that those rebalancing solutions will come. The pressure we feel today from higher oil prices is starting to create incentives for conservation, efficiency, and substitution, and for the development of new process innovations. Visionary companies and individuals will find a new way. Throughout the history of energy, inventors and entrepreneurs like Jacob Rivera and Charles Morgan, James Watt and Thomas Edison, and John D. Rockefeller have made their fortunes by meeting our needs, just in time.

The question remains: how quickly and painlessly can we negotiate that shift now? Alan Greenspan assures us that we have always managed to move on to the next great fuel before the resources available to us have been fully exploited. But he neglects to mention how close we have cut it, and how desperate we have become before the shift was accomplished.

In our thirst for whale oil, for instance, the great sperm whales were nearly slaughtered to extinction. As the whaleships traveled longer distances in search of a more elusive catch, there must have been a sense among the experienced whalemen that an end was coming. Indeed, Herman Melville might have had that emotion in mind when he described the forlorn scene of the remains of a sperm whale unlashed from the side of a whaleship and allowed to return to the sea, "sliding along beneath the surface as before, but, alas! never more to rise and blow."[8] Eventually, the resource that had once seemed so abundant could no longer be found in sufficient quantities to light the world. Somewhere, the last whale oil lamp was lit, and a new energy era began.

Notes

1   A barrel is a standard unit of volume in the oil industry. One barrel is the same as 42 U.S. gallons, 35 imperial gallons, or approximately 159 liters.

2   OPEC is the Organization of Petroleum Exporting Countries, an intergovernmental organization representing eleven of the largest oil producers in the world.

3   In response to Nazi Germany's anecdotal research into atomic weapons, the United States initiated the top-secret and top-priority Manhattan Project in June 1942. Scientists across the country worked on an accelerated agenda to successfully develop the world's first atomic bomb. The non-military spin-off of the Manhattan Project was nuclear power.

4   *Moby-Dick*; or, *The Whale* by Herman Melville p. 536; 1972 Penguin Books, New York.

5   *Moby-Dick*; or, *The Whale* by Herman Melville p. 204; 1972 Penguin Books, New York.

6   Remarks by Chairman Alan Greenspan to the National Italian American Foundation, Washington, D.C., October 15, 2004.

7   See Radical Innovations and Substitutions arrow leading into the evolutionary cycle in Figure 1.1.

8   *Moby-Dick*; or, *The Whale* by Herman Melville p. 537; 1972 Penguin Books, New York.

CHAPTER 2

THE THIRTY-THREE PERCENT ADVANTAGE

Today we have advanced well beyond whale oil and kerosene to a mixture of fuels that meet our needs for illumination, power, and transportation. Those fuels include petroleum, natural gas, nuclear power, coal, and even solar and wind power. Together, they combine in ways we can barely imagine to make our air conditioners work and our factories hum.

While the crucial and irreplaceable fuel is petroleum—the light sweet crude mentioned in Chapter 1—our reliance on the mix of fuels is so seamlessly woven into our daily lives that we take cheap, secure, clean energy almost entirely for granted, a birthright of our modern age. Indeed, it is only during those rare periods when our birthright is threatened—times when the pressure in our energy cycle is building rapidly—that we become concerned about where our fuel comes from and how we can continue to secure it at a low price to preserve our way of life.

Changes in the world of energy are measured not in months, not in years, but often in decades. The abrupt transition from whale oil to kerosene took less than two decades. In the history of energy substitutions, that's a duration of time akin to an eye blink. It's a rare event when we switch from one fuel to another, or even switch to alternative technologies that use the same fuel in different or better ways, and there must be compelling reasons to make the shift. Recasting consumer habits is a large undertaking, but the primary obstacles to real change come from the inflexibility of the technological standards and physical infrastructure that are placed up and down the energy supply chain. For example, our oil-fed energy supply chains have developed over a 145-year-old growth cycle, ever since spermaceti gave way to kerosene. In that time, a massive energy nexus has been solidly welded into every corner of the modern world. We are dependent on this multitrillion dollar global infrastructure as much as we are dependent on the petroleum that feeds the entire supply chain. Is it any wonder that influential nations of our world have, over the past 100 years, sought to secure and control the commodity that underpins our society?

The Conversion of Energy

Simply put, we need energy because of the work it can do for us, and so we have developed elaborate supply chains to obtain that fuel cheaply and reliably. But our world is not only served by those supply chains; it is also shaped by them. Every time we have switched to a new primary fuel, society has undergone some fundamental reorganization as a result. In an ever more interdependent world, these switches have had progressively greater geopolitical overtones.

For 40,000 years, controlled fire was our primary source of energy. We gathered sticks and twigs to stockpile that resource, then burned those sticks and twigs to convert the energy stored in the wood into heat and light. In this way, we cooked meat and warmed our habitats. No doubt the diverse range of human cultures developed as a result, since the availability of light was crucial for creating more time for talking, storytelling, singing, and art making.

Then, around 4000 BC, we discovered how to harness the power of animals. At first, animals were used only for simple tasks like carrying goods and dragging firewood. Our great leap forward into the Agrarian Age occurred when we tied an ox to a wooden beam and led it in tight circles around a well in order to pump water from the ground. The ox-powered pump was a profound technological innovation that revolutionized human life and shifted us to a new primary fuel source with its own energy supply chain. Wood was still gathered for heat and light, but to power pumps and keep fresh water flowing to the fields, we now needed hay. In that sense, not only did the ox-powered pump make agriculture possible, but it also made agriculture imperative.

In terms of that energy supply chain, the picture looked something like this: The primary energy feedstock—found upstream in the supply chain—was the grass grown in the fields. Downstream, that grass was cured and converted into hay, which could then be fed to the ox. The ox-powered pump was the primary energy conversion technology. In essence, the energy in the grass was converted into the energy needed to pump water.

The notion of conversion is a central theme in the story of energy. In 1847, Hermann von Helmholtz, a German physiologist and philosopher with a keen interest in math and physics, postulated one of the most important laws of physics. The First Law of Thermodynamics states that energy may be transferred or converted into different forms, for example, heat, light, and electricity, but it can neither be created nor destroyed. Helmholtz's proposition derived from his recognition that all forms of energy are fundamentally the same. In other words, the energy in the chemical bonds of a substance like whale oil is the same as the mechanical forces rotating a gearbox or the electromagnetic waves found in light.

We say that energy is *transferred* when it goes from one system to another. For example, a gearbox executes a mechanical-to-mechanical energy transfer from one gear to another. Energy may also be *converted* from one system to a different system as when a fuel is burned in a lantern to convert chemical energy into light energy, with heat as the usual by-product.

Helmholtz also stated that energy cannot be created or destroyed. This principle of the *conservation* of energy means that there can be no net gain or loss in energy when it is transferred or converted from one system to another. Nevertheless, real-life transfers and conversions are never pure. Frictional forces in a

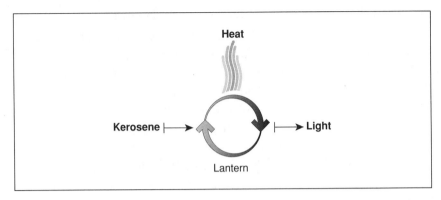

Figure 2.1   Conservation of Energy in a Lantern: Conversion of Kerosene into Heat and Useful Light

gearbox, for example, transform some of the mechanical energy into heat. No energy is created or destroyed in the process, but it can be lost to us, in the sense that it does not get applied 100 percent to the task that we desire of it. Applying more and more of the energy in a primary fuel directly is a problem that physicists and engineers grapple with all the time; their efforts are intensified when the pressure in the energy cycle is approaching a break point.

Of course, the reason why we convert or transfer energy intentionally is because it provides us with the capacity to do work. When energy is transformed from one form to another, say from chemical energy to heat energy, we are able to extract some of the energy and put it to useful work. The ox-powered pump was a conversion device that allowed us to turn the chemical energy of hay into the mechanical energy needed to pump water out of the ground. This is such a useful conversion process that it is still predominant in some parts of the world today.

But in England, around the fourteenth century, a new supply chain began to emerge when heat energy was extracted from the burning of coal. Records suggest that coal was first introduced as a fuel in Scotland in the ninth century by monks to heat their abbeys. Over time, coal power became adopted by brewers and

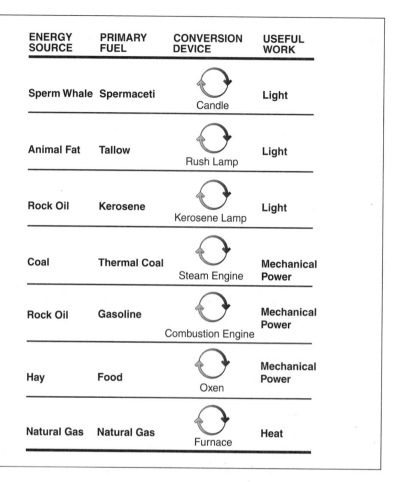

ENERGY SOURCE	PRIMARY FUEL	CONVERSION DEVICE	USEFUL WORK
Sperm Whale	Spermaceti	Candle	Light
Animal Fat	Tallow	Rush Lamp	Light
Rock Oil	Kerosene	Kerosene Lamp	Light
Coal	Thermal Coal	Steam Engine	Mechanical Power
Rock Oil	Gasoline	Combustion Engine	Mechanical Power
Hay	Food	Oxen	Mechanical Power
Natural Gas	Natural Gas	Furnace	Heat

Figure 2.2   Examples of Energy Supply Chains: Primary Conversions into Useful Work

smiths. Demand grew to such an extent that by the fourteenth century, a coal trade had developed in England.

In a sign of things to come for the energy industry, the use of coal was simultaneously encouraged and discouraged by the government. To conserve rapidly diminishing forests needed for shipbuilding, a penalty was applied to those who burned wood to power their smithy or brewery. But the government also prohibited coal burning for a time because of the tremendous pollution it caused. Nevertheless, in part because wood was an increasingly scarce resource, coal became an increasingly compelling option for substitution. In effect, people decided that the utility of coal was so valuable that its pollution was worth putting up with.

By the mid-1600s in England, the burgeoning iron trade increased the demand for coal and put great pressure on the tightening resource. The cheap, accessible coal available near the surface of the ground was rapidly becoming depleted. In order to tap more supply, coal miners did what whalemen would later do when whales became scarce and what oil drillers would later do when light sweet crude oil began to run out: they went to greater lengths to find more of their increasingly precious resource by digging ever deeper pits.

A new challenge soon arose, however, as these deeper pits needed to be drained of water on a nearly constant basis. Although crude pumps existed, the power of these horse-powered machines was very limited. The demand for coal was showing no signs of relenting, so something better was needed if the coal industry were to continue to supply its market.

A radical new technology emerged to save the day when the steam engine was invented. Noisy, dirty, yet effective, this new device was a source of power that helped to pump water out of underground coal mines. The steam engine's usefulness as an energy conversion device was so spectacular that it would be applied to a wide variety of other innovations. In the process, the world switched resolutely from wood to coal, its first major fuel substitution. Of course, nothing was ever the same again.

I Sell What All the World Desires

When it comes to the discovery of steam power, most people immediately think of James Watt. According to legend, he sat mesmerized in the kitchen as a young boy, watching the kettle boil over, ignoring his mother's shrill demands that he do something more useful with his time. In fact, steam power had been exper-imented with since 100 BC. And it was steam that some inventors later turned to as a possible means of powering a pump that would be able to drain water from the deep coal mines of Scotland. It wasn't until 1712, however, in Cornwall that Thomas Newcomen invented a steam engine that could harness the power of steam effectively. Newcomen was an engineer who lived near the coal mines and had learned of the experiments being done to build a steam-powered pump. His own version, the Newcomen Steam Engine, was an improvement on an approach taken by Thomas Savery a few years before. The Newcomen Engine drove a piston that then powered a pump to suck up the water. Strictly speaking, it did not use steam to drive the piston directly. Instead, the steam generated in the engine created a vacuum in a separate boiler, and it was that atmospheric pressure change that forced the piston to move. Nevertheless, the Newcomen Engine could do the work of 40 horses—an impressive and previously unmatched harnessing of power.

The main drawback of Newcomen's engine was that it consumed a tremendous quantity of coal. In a way, it was as though an ox had been found that could pump a bounteous quantity of water into the fields, but needed a distressingly large quantity of hay to feed it. As a result, few of the coal mines could afford to build, let alone operate, a Newcomen Engine, and not many were sold.

A half century later, in 1778, James Watt's steam engine would be widely adopted and was rightly credited with kick-starting the industrial revolution. His engineering brilliance, however, was only part of the story. The other two essential elements behind his steam engine's success were its capitalization and patent pro-tection. Indeed, this three-part equation has been essential to the success of many other significant scientific and industrial devel-opments ever since. In Watt's case, it is even more remarkable that he managed to design, perfect, build, and sell an expensive piece of industrial machinery without any state or institutional financing. Of course, he did have some help—a story that's been under-reported in the history of energy.

James Watt came from a family of mathematicians and ship builders. He had a brilliant engineering mind. When he saw something mechanical, it was not enough for him to understand how it worked, or even how to fix it; he wanted to discover the principles of physics behind the device in order to improve upon the technology and sometimes radically change it. When Watt learned about the Newcomen Steam Engine, the possibilities of steam power immediately captured his imagination. Could steam power be used to drive a carriage on wheels? Could it run a paddle boat? This excitement led Watt to locate a small working model of a Newcomen Steam Engine, and to experiment on it.

Naturally, given his inclination for seeking improvements, he immediately saw some flaws. Newcomen's engine consumed so much coal because it was fundamentally inefficient. A great deal of heat was being lost somewhere in the boiling process. Using the principles of condensation he learned during these experiments, Watt added a third, separate vessel to the boiler and condenser Newcomen used in order to condense steam more efficiently. Next, Watt improved upon the mechanical design. Newcomen's engine used a piston with an up and down motion. Watt decided to adapt it to rotary motion. It may seem obvious today, but at the time the idea of transferring steam energy into rotating mechanical energy was revolutionary. Together, his improvements made Watt's engine three times as efficient as the Newcomen Steam Engine and invented the first powered wheel.

Gaining efficiency, or getting more useful work out of the energy contained in a primary fuel, is a paramount concept in conservation. It's especially important in the face of nonrenewable fuel supplies like oil and natural gas that are getting progressively harder to find. In general, our society's use of energy is dismally wasteful, creating lots of opportunity for the building of a better mousetrap. In many cases only a small percentage of the original energy contained in a primary fuel like crude oil is actually harnessed as useful work. For example, by the time the "rubber hits the road" in our cars, only about 17 percent of the energy in a barrel of oil actually ends up making the wheels go round, much less if you're stuck in traffic. Helmholtz's conservation of energy law is inviolable; all energy must be accounted for. So the remaining 83 percent required to drive from your home to your office is wasted further down the supply chain, with the biggest wastage coming near the end: the notoriously inefficient internal combustion engine. If today James Watt could miraculously make our oil-based transportation supply chain three times more efficient, say 51 percent instead of 17 percent, the world would consume 29 million barrels per day less oil[1], and preserve our oil legacy perhaps for another century. No doubt, the benefits to our environment and climate would be just as profound.

Given such an improvement in efficiency, Watt believed that his approach to the steam engine would have great commercial value—not only for coal mining but for many other industrial uses as well. He had no money, however, to pay for his research, let alone establish the production capacity needed to build and sell the end-product on a profitable scale. He was a man of ideas, not a man of the world. He had been in poor health since youth, and preferred to save his vigor for thinking and tinkering, rather than expend it on business matters. He knew he needed a partner to finance and guide his endeavors. He went through several before he met Matthew Boulton, a man who recognized the magnitude of what Watt had done, and made sure that they both could profit from it.

Boulton was born into wealth, but he was the kind of man who was not satisfied with what he had been given and wanted to turn it into something greater. His family money had come from the hardware business. To grow that business, Boulton established a world-class hardware factory north of Birmingham, calling it the Soho Works. The factory became profitable very quickly, as demand for its high-quality goods exceeded capacity.

James Watt visited the Soho Works in 1767. He was impressed with the precision of Boulton's machines, and with the organization of his factory. Boulton was clearly a man who had a remarkable sense of manufacturing organization, an adeptness at the management of capital, and engineering skill to boot. As an industrialist, Boulton was keenly aware of his own need for power and recognized that Watt's steam engine provided a revolutionary new way to receive power on demand. Boulton envisioned an industrial-scale factory that produced steam engines for sale all over the world. He also had the unique blend of skills necessary to make such a vision into reality.

Enthusiastically, Boulton and Watt formed a partnership in which Boulton would finance Watt's research and development in return for 40 percent of the profits—probably the first recorded private equity deal in the history of the energy industry. By 1778, the Boulton & Watt Company produced its first steam engine. It was not an easy road. The technological hurdles were steep and very costly. Boulton nearly ruined himself financially several times along the way. Moreover, it was difficult to protect the investment from outsiders who were quick to poach upon the ideas. Years of work, not to mention the money and intellectual advances behind those efforts, could be stolen in a short time by someone who figured out Watt's technical innovations and copied them. Boulton recognized this problem from the beginning and lobbied the English parliament for a change in patent law that would extend the length of a patent from 8 years to 25. Watt, meanwhile, worked furiously to come up with more improvements and quickly patent them, in order to leave competitors and pirates behind. The obvious and next hurdle was selling the engines once they were built. At first, they were too expensive for most industrial customers, so a complicated and risky system of financing was devised that would allow coal mines and factories to pay for the engines over time. Customers who at first balked at the price were now able to buy Watt's engines despite the overall cost. Sales quickly followed. The competitive advantage afforded by the steam engine was so considerable that industrialists who wanted to keep up simply didn't have a choice but to buy a steam engine, too.

Figure 2.3   Innovation in Extracting Work from Energy: James Watt's Rotary Wheel Assembly Attached to his Steam Engine
(*Source: The Author's Photograph. London Science Museum, United Kingdom*)

The steam engine had been invented to service the needs of the coal industry, but as the machine was adapted to the cotton mill, the corn mill, the waterworks, the paper mill, the metal industry, and the transportation industry, it was coal that would come to serve the steam engine. In less than a lifetime, the world had converted to coal and its energy supply chain. The steam engine transformed the energy of coal into the power needed by the Industrial Age. As Matthew Boulton said to James Boswell, the famed biographer of Samuel Johnson, "I sell here, sir, what all the world desires to have, Power."

Two hundred years later, our desire for Boulton's sales proposition is stronger than ever.

The Fateful Plunge

The early difficulties that the Boulton & Watt company had in convincing companies to switch to steam engines should resonate with vendors of new energy and power products today. Trying to introduce a technologically superior product into "old school," energy-intense industries is extremely challenging. Capital budgeting decisions are slow and fast payback periods must be demonstrated in order to satisfy impatient shareholders. A device like the steam engine could successfully be introduced today, but only if it has similar compelling utility and economic advantage over our established means of doing work. Even then it would take time. Fuel cells and other energy conversion devices that are being touted for our future—as exotic to us today as steam engines were to the public three centuries ago—simply do not have the same all-around, compelling jump in utility that a steam engine had over a team of horses.

Considering this inherent resistance to change, the transition to Watt's steam engine clearly demonstrates its value. With the advent of this new energy conversion device, England's agrarian society quickly transitioned to an age of coal. Power on demand, provided by the steam engine, catalyzed capitalism. Cottage industries became factories, and those factories became larger and more efficient. Cities grew. The gap in wealth between employer and employee became increasingly vast. Tapping into its suddenly precious coal resources, England became even more dominant globally through trade and commerce. Steam-engine-powered looms, mills, and hardware factories produced goods that were exported all over the world. Because of coal, London was the world's largest, best lit, and most polluted city. Charles Dickens described a late afternoon scene in the grimmest terms: "Smoke lowering down from chimney-pots, making a soft black drizzle, with flakes of soot in it as big as full-grown snow-flakes—gone into mourning, one might imagine, for the death of the sun . . . Gas looming through the fog in diverse places in the street . . . Most of the shops lighted two hours before their time—as the gas seems to know, for it has a haggard and unwilling look."[2] It was a new, though murky dawn.

The success of coal was also remarkable in the transportation industry. America became criss-crossed with rail lines on which steam-powered trains moved people and goods over great distances. Coal trumped wind power too. The British Navy and merchant fleet navigated the globe, connecting the far corners of the British Empire more quickly and efficiently than ever before. Britain and America were both blessed with vast reserves of coal, providing a strong sense of energy security shoring up their displays of economic and military strength. For an island nation like Great Britain, this sense of security was very important. And yet, despite the advantages of coal, and the resulting prosperity that came to Britain, that nation was the first to convert itself to crude oil. The fact that it thought to do so and was able to accomplish the transition in a relatively short period of time shows the pressure that strategic-military considerations can exert on the energy supply chain.

By the end of the nineteenth century, Britain's naval supremacy was under threat because of the rise of an increasingly nationalistic Germany. Since the 1890s, Germany had been pushing for political, strategic, and economic power. In 1897, it began an aggressive drive to build up its navy—a move that was interpreted as a direct challenge to Britain's dominance of the high seas. Talk of this naval race filled the press in both Germany and Britain, creating anxiety among the population and intensifying nationalistic fervor.

If war was unavoidable, as many believed, how should Britain best prepare itself? To John Arbuthnot Fisher, First Sea Lord of the Royal British Navy, the answer had been clear for some time: The British Navy must convert from coal to oil in order to power its fleet.

It was a remarkable belief to hold, but Lord Fisher was a prescient strategist, passionate about the modernization of the British Navy. As early as 1882, Lord Fisher began to preach his cause to the British government, reporting that using oil instead of coal as a fuel would add significant advantage to the value of any fleet. To most British politicians this kind of talk was heresy. British ships were powered by quality Welsh coal, which Britain had in great supply. In contrast, Britain had no oil at all. Moreover, a single American company, Standard Oil, controlled 30 million of the world's 35 million barrels of production per year, almost all of it devoted to kerosene.[3]

Lord Fisher's dream of converting the British Navy to oil seemed grandiose and unsound. Many politicians of his time also opposed Fisher's call to expand the navy. Even young Winston Churchill, although a close friend of the Admiral, joined outspoken Liberal party leader Lloyd George in pushing for a British naval agreement with Germany so that money could be spent on social reforms instead of a naval arms race. But, as so often happens during times of geopolitical tension, a simple event can have far-reaching consequences. In 1911, when a German gunboat sailed into a French colonial port in Morocco, that provocative act set off a political crisis in Europe. Perhaps even more importantly, it changed Winston Churchill's view of Germany. From that point forward, he had no doubt that Germany's aggressive naval expansion was a direct threat to Britain. It could only lead to war, and Britain must prepare itself for that inevitability. At the end of 1911, Churchill was offered the chance to become First Lord of the Admiralty, the top civilian post for the British Royal Navy. He accepted.

Since the gunboat incident, Churchill had bent all his energies toward readying Britain for an eventual military conflict with Germany. Now, as head of the Navy, Churchill faced the choice that had been advocated by Lord Fisher. Should the entire British Navy convert from coal to oil? The shipyards were gearing up for the production of new ships, and a decision needed to be made. On each side of the debate there were significant advantages and potential consequences.

The advantages of oil over coal were many. Oil would provide ships with greater speed, mobility, and radius of action. This would give British battleships a crucial edge to outperform the emerging German fleet. Oil-burning ships could be refueled at sea, even off enemy shores, while coal-powered ships had to be refueled at a base, necessitating that approximately one-third of the fleet would be inactive at any one time. Moreover, unlike coal, oil did not deteriorate during storage. On-board, oil-powered ships required 60 percent fewer personnel for work in the engine and boiler room than coal-powered ships, while using half as much fuel. In the heat of battle, these differences could be deadly.[4] As Churchill noted later, "As a coal ship uses up her coal, increasingly large numbers of men had to be taken, if necessary from the guns, to shovel the coal from remote and inconvenient bunkers to bunkers nearer to the furnaces or to the furnaces themselves, thus weakening the fighting efficiency of the ship at the most critical moment in the battle . . . The use of oil made it possible in every type of vessel to have more gun-power and more speed for less size or cost."[5] Lord Fisher quantified these benefits as providing an oil-powered navy with a 33 percent advantage over a conventional coal-powered one. In his view, this made the verdict clear: "It is a criminal folly to allow another pound of coal on board a fighting ship."[6]

Still, the major disadvantage of oil was so significant that it overwhelmed all of the positives in some people's minds. While large reserves of high-quality coal could be found in Wales, converting to oil meant that Britain would become dependent on importing its most strategic fuel, putting it into direct competition with other nations struggling for their own access. Even if supplies could be secured from other lands, in particular Persia, that oil would still need to be transported by sea to Britain. During wartime, this meant that the nation's strategic supply could be cut off, like an overexposed life line, leaving the homeland helpless. The vulnerability of committing to an extended supply chain made it difficult for many people to see the advantages of oil.

Committing to that supply chain was a risk Churchill decided to take. Under his leadership, the British navy launched three successive major naval programs from 1912 to 1914. In what

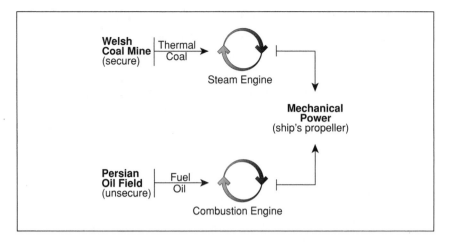

Figure 2.4   Churchill's Strategic Options for Turning the Navy's Propellers: Energy Security versus a Thirty-Three Percent Advantage

Churchill described as a "fateful plunge," the British navy would henceforth be dependent on oil.

Soon after, on June 28, 1914, Archduke Franz Ferdinand of Austria was assassinated in Sarajevo. The complex network of treaties and alliances that had maintained the European balance of power was tripped, and like two opposing domino lines, the countries of Europe fell into war, one after the other. On August 4, 1914, Churchill sent word to the ships of the Royal Navy that they were to commence hostilities against Germany. Over the following bloody years, oil would have the chance to prove its worth more resolutely than even Lord Fisher and Winston Churchill could have imagined.

As Necessary as Blood in the Battles of Tomorrow

There are many examples from World War I in which the supply of oil or lack of supply played a crucial role. The advantages provided by oil- and gasoline-powered vehicles were so clear that the conduct of war would be transformed over the ensuing years. In some ways, it was as though World War I marked the end of one era, with its quaint reliance on ceremony and colorful uniforms, horses, and slow-moving columns of men, to a new century in which the machinery of war would churn violence with unremitting speed and efficiency.

At the beginning of World War I, the British went to France with 827 motorcars and 15 motorcycles. By the end of the war, the British Army had 56,000 trucks, 23,000 motorcars, and 34,000 motorcycles. By January 1915, the British aviation industry had built only 250 planes. During the course of the war, aircraft speed doubled and production figures increased by far more than that. In the war years, Britain produced 55,000 planes; while France produced 68,000, Italy 20,000 and Germany 48,000.[7]

Forty-five percent of the British Navy was dependent on oil. New motor cars, trucks, tanks, and planes were gobbling up the supply of diesel and gasoline. With such a mechanical buildup, it should not be surprising that access to oil was the fulcrum on which the war turned. Britain, through its expertise at tapping international resources, had secured controlling oil interests in Romania, Russia, California, Trinidad, the Dutch West Indies, and the major oil fields of Mesopotamia and Persia. Even so, as the war ground on, the pressure on the Allies' oil reserves intensified, and the fears of those who understood the vulnerability of the supply chain were soon realized. Germany was cut off from its oil supplies. Desperate to level the playing field, it began a submarine campaign to sink and destroy Allied shipping and oil tankers. By 1917, the British were on the verge of a naval oil shortage. The French were in equally dire straits. As French Premier Clemenceau pleaded to U.S. President Woodrow Wilson, "If the Allies do not wish to lose the war, then, at the moment of the great German offensive, they must not let France lack the petrol which is as necessary as blood in the battles of tomorrow."[8]

The entry of the United States into the war helped tip the balance for the Allies, not least in part because of the tremendous oil reserves that became available to power the war effort. In that sense, Lord Fisher and Winston Churchill had been prophetic about the advantages in switching from coal to oil. Despite the vulnerability of the supply chain, the utility of oil was so compelling that it provided the military advantage necessary to win the war. Indeed, as Lord Curzon later famously noted, "The allies floated to victory on a wave of oil."

He Who Owns the Oil Will Own the World

In the postwar period, the lesson of victory through oil wasn't lost on any of the former combatants, least of all Great Britain. The French industrialist and senator Henri Berenger put it bluntly when he said: "He who owns the oil will own the world, for he will own the sea by means of the heavy oils, the air by means of the ultra refined oils, and the land by means of the petrol and the illuminating oils. And in addition to these he will rule his fellow men in an economic sense, by reason of the fantastic wealth he will derive from oil—the wonderful substance which is more sought after and more precious today than gold itself."[9] By extension, as Matthew Boulton had noted during the coal age, all the world now desired control over this new source of energy and its associated supply chain. In the wake of the geopolitical shake-up of World War I, and the growth of commercial oil applications, a great scramble began for the world's unclaimed oil reserves.

The stakes were high. Despite its dominance in domestic production, the United States had less than 12 percent of the world's oil reserves within its own territory. Great Britain had only 6 percent within the borders of its extensive empire. In fact, 70 percent of the world's oil was located in nations and regions (like Russia, Mexico, Venezuela, and the Middle East) whose then political or military weakness invited incursions from outside influences.

The struggle for control over these oil-rich lands has played hot and cold for the last 100 years. Today, for instance, hardly a day goes by when we don't hear news about Mosul, a major northern city in Iraq. Conflict over the region surrounding Mosul—the Transcaucasus to the north, Iran to the east, and Arabia to the south—goes back to the post-World War I era. Immediately following the Great War, Mosul became the focus of intense global attention because it sat above an oil field of tremendous promise in what was then known as Mesopotamia. Oil in the general region was old news. While following the Silk Route to China in 1271, Marco Polo observed that: "On the confines (of Armenia) towards Zorziana (Georgia) there is a fountain from which oil springs in great abundance, insomuch that a hundred shiploads might be taken from it at one time. This oil is not good to use with food, but it is good to burn, and is also used to anoint camels that have the mange. People come from vast distances to fetch it, for in all the countries round about they have no other oil."[10] Over 700 years later we still come vast distances to fetch the oil of the region, the difference now is that we have discovered uses for this oil that Marco Polo never could have dreamed possible.

Mesopotamia had been held by the Ottoman Turks for four centuries. Now with Germany and its Turkish ally defeated in World War I, the British and French began maneuvering in that part of the world for influence. At the heart of their mutual interest was a desire to determine how best to split up the oil of the Middle East.

Initially, in 1916, the British and French cut a deal through the informal Sykes-Picot agreement whereby the British agreed to support French claims to Mosul in exchange for French support to their claims in the Near East. Upon learning about these terms, others in the British government more attuned to the strategic importance of oil raised an outcry over the surrender of such a valuable resource. The British immediately began backpedaling on its own agreement with the French.

Mosul was still officially under Turkish control when the armistice for World War I was signed, but the British pushed forward and captured the city anyway. Conflict then broke out over whether Mosul belonged to Turkey or should be included within the borders of the newly created Iraq, which was now part of the British sphere of influence. France, too, had problems with the British grab of Mosul because of the Skyes-Picot agreement. The British and French began arguing over how far to extend the eastern borders of Syria, which was in the French sphere of influence. Finally, they settled their differences in the San Remo Agreement of 1920. Under that agreement's terms, the British would retain control over Mosul, while the French would get a 25 percent interest in the British-controlled Turkish Petroleum Company in exchange for allowing pipelines to be built across French-controlled Syria. Pipeline access through Syria was imperative for transporting British-controlled oil in Iran and Iraq to a Mediterranean port.

As the agreement was still being formulated, leadership in the United States had grown increasingly alarmed. Charging the British and French with collusion in a conspiracy to block the United States from Mosul, the U.S. State Department demanded that Britain establish an "open-door" policy in the Middle East, which Britain countered with claims of U.S. hypocrisy in Latin America and Mexico. Still, the United States was not comforted by the words of Sir Edward Mackay Edgar, a British petroleum banker, who had put the matter arrogantly, if unwisely, in 1919 when he derided the United States for squandering its own oil reserves and failing to secure new reserves in other regions of the world. As Sir Edward declared: ". . . the United States finds her chief source of domestic supply beginning to dry up and a time approaching when instead of ruling the oil market of the world she will have to compete with other countries for her share of the crude product. The British position is impregnable. All the known oil fields, all the likely or probable oil fields, outside of the United States itself, are in British hands or under British management or control, or financed by British capital."[11]

Though filled with hyperbole, there was still some truth to the criticism and the prediction. By 1928, Great Britain managed to assume control over 75 percent of the world's oil reserves outside the United States. How had the United States, whose oil had saved the allies in World War I, found itself so outmaneuvered in the global chess game just 10 years later? The root of the problem stemmed from the differences in British and American foreign policy as well as in how British and American oil companies were owned and operated. The British believed that oil was such a strategic-military necessity that its procurement demanded government involvement and support. The Americans, in particular the domestic oil companies, believed that government had no place in business. Oil executives aided by their lobbyists were fiercely protectionist, and being nobody's fool, they knew that any U.S. state-sponsored Middle Eastern oil making its way to the lower-48 would harm the market for domestic crude being pumped out of Pennsylvania, Texas, and Oklahoma. In many respects, this made it necessary for the handful of American oil companies active internationally to be the *de facto* arm of the U.S. government in the great scramble for oil that followed World War I, a choice that would have lasting strategic consequences for America.

An Open and Shut Door

In the age of kerosene, the oil industry was dominated by only a few independent companies. The behemoth was Rockefeller's Standard Oil, which managed to outmaneuver most American competitors in selling kerosene domestically and around the world. Although competitors like Gulf and Texaco became formidable rivals, the greatest threat to Standard Oil's global dominance did not emerge from America, but from Great Britain.

Marcus Samuel, a merchant based in London, had an international outlook on commerce and trade. He inherited a small fortune from his father, who had made his money importing shell boxes from the Far East into Britain. When Samuel expanded the family firm's trade in East Asia, he turned to coal as a commodity, distributing it from a base in Japan. Later, after the Czar opened Russian oil reserves to international development, Samuel joined a group (including the French Rothschilds and the Swedish Nobel brothers) that bought and sold that oil. In Europe and America,

Standard Oil had established a dominant position as market leader. But in East Asia, Marcus Samuel saw an opportunity to break that stranglehold and grow his own firm's business. In his secret Far East strategy, he built storage facilities along key distribution centers, and then designed a fleet of tankers capable of passing through the British-controlled Suez Canal. In homage to his father's shell box importing business, each of those tankers was named after a shell—and Shell would end up being the name of Samuel's company. In 1893, the first vessel in his fleet passed through the Suez Canal with Russian oil bound for Singapore and Bangkok.

In dealing with rivals and upstarts, Standard Oil's strategy had always been to lower the price of product in threatened markets to such an extent that competitors went out of business or allowed themselves to be bought. This strategy was viable because Standard could outlast a price war by relying on higher revenues from more secure markets. Marcus Samuel's network of ships and distribution centers and his secure source of Russian oil allowed him to survive and resist Standard's offer of acquisition.

Over the next decade, Standard Oil continued to try to acquire Shell. Rather than accept such a fate, Samuel joined a strategic alliance with a smaller rival based in the Far East, called Royal Dutch. Henri Deterding, a Dutch bookkeeper who displayed a mastery of finance and operational systems, had become the leader of Royal Dutch by the time of the alliance. "Napoleon" Deterding, as he was called, soon dominated the partnership with Samuel. When Shell, during a vulnerable period, became weakened by a renewed price-cutting onslaught by Standard Oil, Samuel was forced to negotiate an unequal merger with Deterding. The new company, formed in 1906, was known as Royal Dutch/Shell.

For many years, Marcus Samuel had been urging Admiral Fisher to convert the British Navy from coal to oil. Of course, it was self-serving for Samuel's Shell, but by all accounts Samuel had best intentions in mind for the British Empire. In turn, it was Admiral Fisher who encouraged Winston Churchill to discuss such matters with Marcus Samuel and Henri Deterding. Churchill was fascinated by Deterding but resisted becoming too closely allied with his company. After all, since its merger, Shell could no longer be relied upon as a loyal, trustworthy agent of the British Empire. Fisher believed that the concerns over foreign influence in the company could be solved by knighting Deterding and making him a British subject. To Churchill, whether Deterding was British or not didn't really matter. The larger issue was that Great Britain had no secure influence over a privately held company.

Standard Oil had been broken up by the American government in 1911 to thwart its monopolization of the markets. But Standard's imprint was still so great that Churchill was able to use the threat of its, and Royal Dutch/Shell's, potential control over oil prices as sufficient argument for getting the British government into the oil business. Under Churchill's direction, the British purchased a 51 percent controlling interest in an oil company called Anglo-Persian, later known as Anglo-Iranian, then British Petroleum, and now BP. The deal was concluded a mere three months before the start of World War I and provided Churchill with the leverage he believed he needed to secure favorable oil prices for the British Navy.

With the conversion of the British Navy from coal, and the commencement of the war, the age of kerosene had given away to the next phase in the crude oil story, the age of naval fuel. Ownership in Anglo-Persian gave the British a leg up in the great scramble for oil after World War I. The company had been formed originally to develop oil reserves in Iran. Its expertise in that region and the support of the British government would give it an advantage in securing Middle East concessions. The main drawback was that a nationally owned company doing business in other countries implied direct foreign meddling. This inspired nationalistic responses in turn by rival nations and by oil-producing states. The United States, for example, was alarmed by the aggressiveness of Royal Dutch/Shell, still British to them, which held concessions in Central America and Mexico, considered very much within the United States' special strategic purvey. As a result, in 1920, a bill was introduced to the U.S. Senate to create the "U.S. Oil Corporation," a company that would be charged with obtaining strategic concessions around the world through the support of U.S. diplomacy. Resistance to the idea of a nationalized business was strong in U.S. political culture, however, and strong opposition by the domestic oil lobby ensured that the proposal failed. Nevertheless, the country that had produced 60 percent of the world's oil and controlled 85 percent of the world's refineries before World War I had finally woken up to the threat that it was being shut out of the larger global supply.

An unfortunate judgment call in Russia by Walter Teagle, Standard of New Jersey's CEO, underscored the fact that the United States wasn't making any gains on the world stage. Opportunistically trying to tap into Russia's share of global oil supply, Teagle ploughed $11.5 million dollars into the Caucasus by buying shares in an oil company run by Sweden's Nobel brothers that was active in the prolific Baku region. But this was in 1920, two years after the revolutionary Bolsheviks had nationalized the oil industry and all its associated concessions. Shell's Deterding, too, was active in buying notionally worthless shares from Tsarist Russian companies. The misguided hope of the fiercely competitive American and British companies was that the Bolsheviks would roll over and the prerevolutionary concession agreements would be honored. In reality, neither could afford not to gamble, lest the Bolsheviks did collapse. Certainly the Americans had to be there. The *London Financial Times* must have made more than a few U.S. policy wonks squirm when they brashly stated, "The oil industry of Russia liberally financed and properly organized under British auspices would, in itself, be a valuable asset to the Empire. . . . A golden opportunity offers itself to the British government to exercise a powerful influence upon the immense production of the Grosni, Baku and TransCaspian fields."[12]

But buying oil assets under fresh communist control was a bad gamble for all parties. At an economic conference in Genoa, Italy, the now-firmly rooted Red Russians refused demands to denationalize. Effectively, all prior concession agreements were worthless. However, the new Russian regime was still open for business, willing to start negotiating fresh agreements. It was a signal that sparked a renewed rush by Standard of New Jersey, Shell, and many other foreign national and foreign independent oil companies seeking to control Russia's oil riches in the southern Caucasus. As someone who followed the political wrangling in 1926, Louis Fischer noted that: "When victory in the World War failed to give the Russian oil prize to any of the Allied nations, they fell to quarrelling for it among themselves. Great Britain, France, Belgium and the United States, these were the factions in the peace-time scramble."[13]

Ultimately, by virtue of historical agreements between Russia, Persia, and Britain, the mess spilled over into the Middle East. Clarity of control didn't really emerge until after World War II, yet even today a sense of permanence still eludes the region.

Whether in Russia, the Middle East, or other corners of the globe, it was left to American oil companies to continue acting on their own behalf in obtaining a share of global oil concessions, and they were late out of the starting gate. Of course, tapping those concessions was in the best interests of those American companies—without a reliable supply of the world's cheapest crude oil, they would be at a formidable disadvantage against competitors like Royal Dutch/Shell and British Petroleum. Still, it is interesting to note the degree to which those companies acted as instruments or proxies of U.S. foreign and military policy. They negotiated like diplomats with the emerging Middle Eastern nations, acted as go-betweens for the U.S. government, and nudged or cajoled the U.S. leadership into action when more official statesmanship, pressure, or threat was required, all while apprising the U.S. government of its own strategic, diplomatic, and military interests. They became experts in a region of the world, and of a strategic commodity, that the U.S. leaders seemed to have surprisingly little interest in securing . . . until that lack of security proved threatening.

The 1920 San Remo agreement between Britain and France divvied up oil concessions in the Middle East, Asia Minor, Romania and French and British colonies. In Mesopotamia (now Iraq) the instrument of this oil development was the Turkish Petroleum Company. Standard Oil of New Jersey was the first American company to decry being shut out of Middle East oil by this "closed door" policy. It was through the insistence of Walter Teagle and the clamoring of a number of other American companies that the U.S. government began to exert pressure on Britain and France to open that closed door.

Those negotiations would take several arduous years. Through the prodding of the U.S. government, an initial agreement was struck in 1922, which gave American oil companies—Standard of New Jersey (later to be Exxon) and Socony Vacuum (later to be Mobil)—a toe-hold in the Middle East through a 20 percent interest in the Turkish Petroleum Company. In 1925, the Turkish Petroleum Company was renamed the Iraq Petroleum Company after the Iraqi government formally granted it a concession. The name change was largely a gesture, because the company's bylaws insisted that the company be British and the Chairman be a British citizen. Further, no agreements between local leaders and foreign oil companies could be finalized without British approval. Effectively, this gave the British veto power over any and all exploration and development. For the Americans, however, it was a foot in the crack of a narrowly opened door.

After all the shares were divided up, the Iraq Petroleum Company was 50 percent owned by Anglo-Persian, 23.75 percent owned jointly and equally by Exxon and Mobil, with the rest split between Royal Dutch/Shell, a French consortium, and Calouste Gulbenkian, a shrewd businessman who was an original founder of the Turkish Petroleum Company. Gulbenkian would later become known as "Mr. Five-Percent," because of his personal equity interest of the same amount in IPC. Aside from becoming one of the wealthiest individuals of his time, Gulbenkian was hugely influential in the geopoliticking. His insistence that none of the partners in the Iraq Petroleum Company seek concessions in the former Ottoman empire—a leftover clause from the Turkish Petroleum Company—was a pivotal event in shaping the geopolitical landscape for oil. According to legend, Gulbenkian simply drew a red line on a map, effectively encircling a giant zone of noncompetition between the shareholders, defining the breadth of the Ottoman Empire, and giving the famous "Red Line Agreement" its name.

Figure 2.5   Historic Boundary of the Red Line Agreement: Mapped onto Today's Political Borders

The signing of the Red Line Agreement in 1928 ended the negotiations over how the post-World War I Ottoman oil riches were to be carved up. Although the British had grudgingly given the assertive Americans the "open door" that they wanted, the unintended outcome was that two of the largest American oil companies were effectively locked in a house, unable to independently seek out the riches of the unexplored deserts in the rest of the Middle East, particularly Saudi Arabia.

Standard Oil of California (later to be Chevron) was not part of the Iraq Petroleum Company and felt no restrictions imposed by the Red Line. Consequently, it established its first concession in the Middle East on Bahrain Island, just off the coast of Saudi Arabia. Meanwhile, Gulf Oil had been offered concessions in Saudi Arabia but could not accept them because of its membership in IPC. Gulf transferred its concession to Standard of California, which then negotiated to secure 200 million acres in Saudi Arabia for exploration.

Meanwhile, the British made a poor political move in Saudi Arabia that would lock them out of that country, red line or no red line. By backing the Hashemite Kings in their war against the victorious Wahabi tribes of Ibn Saud in the 1930s, the Saudi's henceforth would not look kindly upon any British participation in developing (and certainly not controlling) the Kingdom's oil riches. The door was opened wide for U.S. interests, and they made the most of their *entrée*.

The American geologists who arrived in Saudi Arabia must have felt like the whalemen of their day, traveling to the far reaches of the world to bring home needed oil. To downplay their presence among the locals, they grew beards and wore Arabian clothes while suffering in 125-degree Fahrenheit heat and relying on primitive facilities. Their early drilling attempts were disappointing. To minimize the risk and share the costs, Standard of California sold a 50 percent interest in its concession to The Texas Company (later to be Texaco). The new joint venture was called the California-Arabian Standard Oil Company, or Calarabian, and would later be named Aramco. It wasn't until 1938 that Calarabian drilled deep enough to discover oil in commercial quantities, just before the world was plunged into war again.

The American Age

Securing the oil supply chain would be crucial in World War II, just as it had been in World War I. The Germans and British would fight for oil in the Middle East, repeating battles from World War

I throughout the old Ottoman Empire. The Germans would also strive to capture the vast oil fields in Romania, while Russia's oil supplies gave it a strategic advantage against the Allies and then the Germans. Knowing that oil had been the "blood of victory" in World War I, German scientists developed an expensive process to turn coal—of which it had vast reserves—into gasoline and other petroleum products.

In the war in the Pacific, Japan's early incursions into Indonesia and Singapore were similarly geared toward securing a strategic supply of petroleum and other natural resources. As an island nation with no reserves of its own, it was and still is deeply sensitive to any threats to its energy supply. One Japanese justification for the attack on Pearl Harbor was that the United States had been slowly strangling its supply of oil, an aggressive blockade strategy that was tantamount to war. Nevertheless, it is strange that at Pearl Harbor, Japan did not launch a third wave of aircraft to destroy the oil tankers that stored much of America's own Pacific reserves, because damaging that supply chain might have severely hampered American war efforts. In any event, the Japanese reaction highlights the extreme sensitivity that nations have towards their energy security, especially when their energy supply chain is put under military pressure.

In the Middle East, British preoccupation with the Second World War prevented any significant development of reserves. But the U.S. government, viewing operations in Saudi Arabia, Kuwait, and Bahrain as strategically crucial in a time when domestic production was being mopped up by growing demand, encouraged American oil companies to invest heavily in production and refining facilities. The directive was clearly spelled out by Charles B. Rayner, Petroleum Adviser to the State Department, who in a report published on February 10, 1944, noted that: "The Department of State has, therefore, taken the position that the public interest of the United States requires maximum conservation of domestic and nearby reserves and large-scale expansion of holding in foreign oil reserves by United States nationals. It has, therefore, actively supported the efforts of United States petroleum interests to secure and to consolidate concessions abroad."[14] The scramble for oil begun 25 years earlier was still on, this time with the U.S. government fully involved.

At first blush, it seems the Americans wanted control over Saudi oil to fuel the U.S. military complex and domestic commercial markets. In fact, that wasn't really the original intent, as this 1945 memo excerpt from the U.S. government suggests: "the Navy wants Arabian oil developed to supply European commercial demand, replacing western hemisphere oil which might otherwise go to Europe, thus conserving supplies which are subject to U.S. military control."[15]

Clearly, the U.S. Navy valued domestic oil for national security, much as the British valued Welsh coal for their navy 40 years earlier. The memo continues, "Obviously, this concept will never be popular with the American petroleum industry, other than the two companies (Texas and SOCAL) interested in Arabia." As we'll see later, this would not be the last time that strategic military interests would compete with the public's interest.

The significant problem for the Arabian-American Oil Company (or Aramco) was how to transport that oil across 1,500 miles of desert to the eastern Mediterranean. Because of the risk, the cost, and the strategic value of getting the oil to market, it seemed a worthy exception to the aversion for direct U.S. government involvement in the oil business. An initiative was begun to build a U.S. funded oil pipeline from eastern Saudi Arabia to a Mediterranean port.

Harold Ickes, President Roosevelt's highly effective Secretary of the Interior, took the lead in the proposal. He had the approval of the President, the State Department, the War Department, the Navy Department, the Joint Chiefs of Staff, and the Army and Navy Petroleum Board. Under the terms he negotiated, the U.S. government would build the pipeline and requisite facilities, and only charge enough in usage fees to cover the maintenance, operating, and loan costs. In return, the American companies, led by Standard Oil of California, would maintain a one-billion-barrel crude oil reserve for the U.S. military, and provide the U.S. government with the option to purchase that oil at a 25 percent discount.

It seemed like a good and sensible deal to all the involved parties, but not to the free-market American business interests. Once again, domestically bound American oil companies were against any sort of government intervention that would benefit select international companies within the industry. Their primary complaint was that the deal would provide Aramco with a competitive advantage by giving it subsidized access to cheap oil, but a different set of arguments was used to rile the Public, the Media, and Congress. If the U.S. government invested so heavily in oil development in the Middle East, that meant less investment in domestic oil development. Moreover, it would entangle the U.S. government in that turbulent part of the world for generations. Finally, it was argued that the U.S. government simply shouldn't be interfering in business.

In the face of such criticism, the plan to subsidize the pipeline was abandoned as politically unviable. Aramco would end up building the pipeline anyway, with a consortium of American companies. Because of the way American companies had maneuvered to control the Saudi concessions, the British were shut out even more resolutely than the Americans had been 20 years before. As a result of the pipeline, which was a keystone event in bringing Saudi oil to western markets, a special relationship was fostered between Saudi Arabia and the United States. And because of that relationship, which remains in place today, the oil that was once intended to facilitate U.S. energy security by pushing it onto the Europeans has become a hotly debated source of dependency for the U.S. energy complex today.

The Remake

The seemingly unbreakable British dominance of oil in the Middle East had finally been cracked. The United States, without the benefit of a national oil company, had gone from a nation with little influence on the world's oil stage at the start of the century, to a major controller of the world's foreign production by midcentury. The percentage of the world's oil reserves under the

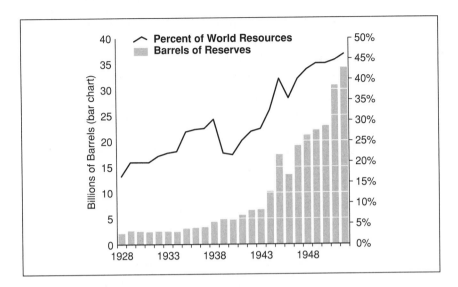

Figure 2.6   Share of the World's Oil Reserves under U.S. Control, 1928-1953: Total Volume and Percent of World Reserves (*Source: Adapted from Fanning, L.M. Foreign Oil in the Free World, p. 359*)

control of U.S.-based oil companies—mainly five of the seven sisters: Exxon, Chevron, Mobil, Gulf, and Texaco—grew rapidly after World War II (see Figure 2.6).

Lasting 25 years, the post-World War I scramble for oil reserves was the twentieth century's first cold war. That the United States and Great Britain were rivals in this conflict is surprising, given that the nations fought together in the two world wars. But the stakes were enormous and the threat to each nation's security was very real, for the age of naval fuel had expanded to a new age where gasoline, jet fuel, asphalt, and a myriad of other petroleum products were necessary to power a military with global reach. Above all, this not-well-recognized cold war implicitly demonstrated that major nations of the world were now dependent militarily and commercially on a substance that ostensibly started out as a humble savior for the whales.

Though 75 years of growth in consumption had bred dependency, the post-Second World War reliance on oil was becoming a lesser concern to many nations, above all the Americans. The public developed a mind-set that was subliminally reinforced by the trend of American oil dominance shown in Figure 2.6. Any concerns about oil were mitigated by a feeling of security imparted by the influence of a handful of the largest corporations on the earth, the giants of the American oil industry that had successfully tied up the world's oil supply lines.

Yet these subliminal feelings of energy security only drew upon half the story. The other half of the story, still largely ignored today, is that foreign imports of oil by the United States have been creeping up steadily over the past 30 years, especially the last 20. The country that dominated production and exports between 1859 and 1900 has been growing a production deficit that now runs close to 13 million barrels per day.[16] Expressed as a percentage of all crude oil consumed, Figure 2.7 shows how American dependence on imports has grown from 10 percent in 1970, to 65 percent by

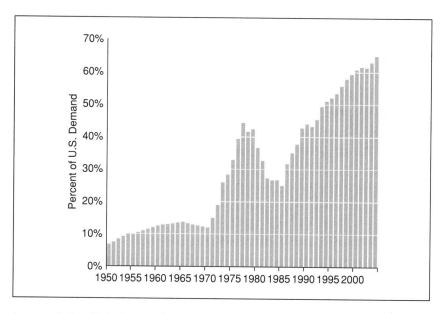

Figure 2.7   U.S. Dependence on Foreign Oil Imports, 1950-2004: Percent of Demand Fulfilled by Foreign Imports (*Source: Adapted from U.S. Energy Information Agency data*)

the end of 2004. Rebalancing with nuclear power and coal helped ease the dependency temporarily in the late 1970s and early 1980s, however the effect on displacing foreign oil was short lived.

At the current rate of unchecked import growth, Americans will be between 70 and 75 percent reliant on foreign oil by the middle of the next decade. And that's not all. Whereas American-based oil companies produced 45 percent of foreign oil in the 1950s[17], that share has dropped down to about ten percent today[18]. The United States is now more dependent and less secure than ever, just as China has recently emerged as America's competition in the great oil scramble of the new century.

Geopolitical tensions have defined those moments of pressure in the energy cycle in the past. In his history of the oil conflict of the early 1900s, Ludlow Denny wrote in 1928 with a poetic world-weariness in describing how those tensions arose again and again. "And the struggle for oil goes on, menacing this flimsy peace."[19] Given the extent of our reliance on cheap energy now, we shouldn't expect the story to be any different for us in the future.

Notes

1 Calculation assumes 50 percent of every barrel of oil today goes toward road transportation and 50 percent goes to other markets. The calculation also assumes the same total distances are traveled before and after the efficiency gain.

2 *Bleak House* by Charles Dickens, p. 2; 1991, Oxford University Press, New York.

3 *We Fight for Oil* by Ludlow Denny, p. 24; 1928 Alfred A. Knopf, New York.

4 *We Fight for Oil* by Ludlow Denny, p. 24-25; 1928 Alfred A. Knopf, New York.

5 *The Prize: The Epic Quest for Oil, Money and Power* by Daniel Yergin, p. 156; 1991, Simon & Schuster, New York.

6 *We Fight for Oil* by Ludlow Denny, p. 24-25; 1928 Alfred A. Knopf, New York.

7   *The Prize: The Epic Quest for Oil, Money and Power* by Daniel Yergin; 1991, Simon & Schuster, New York.
8   *We Fight for Oil* by Ludlow Denny, p. 27-28; 1928 Alfred A. Knopf, New York.
9   *We Fight for Oil* by Ludlow Denny, p.16; 1928 Alfred A. Knopf, New York.
10 Forbes, R. J., *Studies in Early Petroleum History*, E. J. Brill, Netherlands, 1958, page 155.
11 *We Fight for Oil* by Ludlow Denny, p. 18; 1928 Alfred A. Knopf, New York.
12 *London Financial News*, December 24, 1918.
13 *Oil Imperialism: The International Struggle for Petroleum* by Louis Fisher; 1926 International Publishers, New York.
14 *The Secret History of the Oil Companies in the Middle East*, Volume I.
15 *The Secret History of the Oil Companies in the Middle East*, Volume I.
16 Extrapolated from BP Statistical Review 2004 data; the figure 13 million barrels per day includes imports of petroleum products like gasoline. Crude oil imports alone were approximately 10.5 million barrels per day in 2005.
17 Fanning, *Foreign Oil and the Free World*, page 352.
18 US Energy Information Agency Financial Reporting System.
19 *We Fight for Oil* by Ludlow Denny, p. 15; 1928 Alfred A. Knopf, New York.

CHAPTER 3

NOT A WHEEL TURNS

As usual, President Richard Nixon put it bluntly, no doubt igniting the righteousness of some while confirming the views of others. In 1973, at the height of the nation's energy crisis, he stated publicly: "There are only seven percent of the people of the world living in the United States, and we use thirty percent of all the energy. That isn't bad; that is good. That means we are the richest, strongest people in the world and that we have the highest standard of living in the world. That is why we need so much energy, and may it always be that way."[1]

Whatever feelings Nixon's words might inspire in you, the facts he related speak to the success of the American economy and its unique position as an energy consumer. Thirty-two years later, those numbers have stayed roughly the same. In the community of nations, the United States remains the largest energy consumer on Earth. For the last 50 years, America has had no sizable competition for the world's energy resources. But today, as China awakens with its own rapidly growing energy needs, the tension over the global energy supply is mounting.

In his comments, Nixon recognized a law of energy dependence, as fundamental economically as the physical laws formulated by Helmholtz to describe thermodynamics. In every historical example of energy supply, the better and more robust a fuel is, the more we put it to work in our daily lives. In turn, the more successful a fuel is, the more necessary it becomes to the well-being of the overall economy. This creates a dependency that grows deeply rooted over time, and is readily fed as long as cheap supply of energy is available. Once pressure is put on the supply and demand balance, the dependency begins to take on all of the characteristics of an addiction, including the financial hardships and consternation that all addicts experience as they begin to lose control of servicing their habit.

Consider the early history of crude oil. Militarily, the 33 percent advantage was crucial for victory in World War I, compelling the British naval switchover from coal. Economically, the utility of oil was equally compelling for society, transforming every aspect of daily life. From a business perspective, World War I demonstrated that Colonel Drake's Pennsylvania Rock Oil—which up to the early 1900s was still primarily being used as an illuminant in kerosene lamps—had successfully penetrated a new market in the form of naval fuel. By 1920, crude oil had turned out to be a marketer's dream: a core platform product that could be leveraged into multiple markets—aviation fuel for airplanes, gasoline for automobiles, diesel fuel for trains, fuel oil for factories and power plants, asphalt for roads, lubricants for machinery, and even petrochemicals for candles and plastics.

Looking around at the world today, the entrenchment of crude oil parallels how extensively the personal computer has penetrated a broad cross-section of society, business, and government, and become embraced as a new essential by individual consumers. Such platform products don't come around often, but when they do, visionary entrepreneurs like John D. Rockefeller and Bill Gates can become titans of the new industry, and opportunities are rich for investors. One of the more prescient early historians of the oil industry was an American investor named Reid Sayers McBeth. In his 1919 book, *Oil: The New Monarch of Motion*, McBeth observed the great changes taking place in America and wrote that, "Petroleum today holds the front of the stage in a greater degree than ever before. As a wealth creator it never has been so fruitful as at present."[2] The reason for McBeth's bullishness was simple: He saw that average consumers were starting to buy gasoline-needy cars. Noting that ships, airplanes, and nearly everything industrial was becoming increasingly dependent on oil products, he stated that, "Not a wheel turns, which is not dependent on petroleum."

In hindsight, McBeth's observations seem obvious, drawn as they were on the brink of an era of fantastic change. In the world at the time, there were 1.8 billion people, many of whom happily consumed oil products to turn the wheels of their own growing prosperity. The tidal wave of transformation reshaped our lives and environment even more extensively than coal. It was an inventor's dream, a business man's dream, an investor's dream, and a consumer's dream. If that dream has an element of nightmare to it today, it is the extent to which we have become addicted to something that is no longer quite as easy to obtain as in decades gone by. Economic growth and energy use go hand in hand. The intensity of our need for a single core product is troublesome, however, for simple reasons. One product leading into multiple markets seems great when looking down a distribution chain that fans out into every corner of society. But look backwards up the chain, and you see a funnel narrowing to a bottleneck through which that one product must flow. In the face of competition for a limited supply of energy, great effort goes into husbanding the resource, protecting it, and ensuring its security to feed the addiction. This is no different today, in our emerging multipolar energy age, than it was in the great scramble after World War I.

Shaking an addiction, even by replacing it with another substitute addiction, is never easy. It's painful. It often requires a wrenching social or government response. It generally inspires a chaotic flurry of innovation and enterprise. And it takes time. Knowing why and how that substitution occurs will make the journey more understandable, more manageable—and very profitable for those who anticipate and negotiate the coming changes successfully.

Barreling Down the Track

In macroeconomic terms, McBeth sensed that every corner of the economy was being fueled by petroleum. Naturally, the reverse of this observation was also true: petroleum demand was being fueled

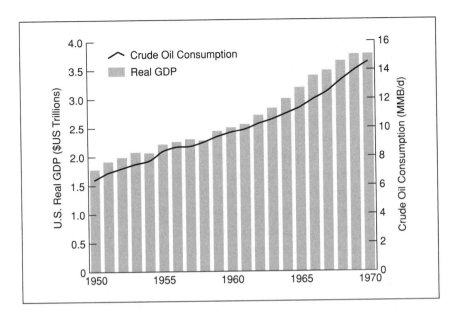

Figure 3.1   U.S. Oil Consumption and Real Gross Domestic Product, 1950-1970: GDP Inflation Adjusted to 2004 Dollars (*Source: Adapted from U.S. Energy Information Agency data*)

by the growing economy. This hand-in-hand relationship is captured in Figure 3.1, which is a graph that plots the high-growth years (1950 to 1970) of U.S. oil consumption and inflation-adjusted U.S. gross domestic product (GDP).

Note that the economy and oil consumption grew proportionally. The proportional relationship is rich with meaning, too, but I will save that discussion for Chapter 4. Suffice it to say, a tight symbiotic relationship between oil and the growth of the economy emerged after World War I and carried through the glory years of western industrialization. Even though the United States was one of the world's largest oil producers, anyone looking at this graph can easily see that securing oil outside its own domestic production would emerge as a crucial necessity over the course of the century.

By the end of World War II, the United States had staked its claim to many of the most important major oil producing regions of the world, and established its special relationship with Saudi

Arabia. In this endeavor, it no longer had any rivals. Britain and France had waned as world powers. West Germany and Japan had no military and were reliant on the United States in extraterritorial matters. While many think of the post-World War II era as being defined by the Cold War between the United States and the U.S.S.R., that military and ideological conflict did not extend into the purview of oil. Russia, after all, had long been self-sufficient in oil, effectively leaving the United States unimpeded in its own efforts to corner the market on global supply.

More cars. More airplanes. More factories and fuel-heated homes. The growth of the American economy in the 1950s and the roaring 1960s was fueled by oil. To get a sense of how petroleum-powered applications made inroads in the economy over that time, consider the railway industry and the conversion from coal-fired to diesel-powered locomotion. Most people are surprised to learn that the steam locomotive was still being produced commercially as late as the 1960s. In fact, it took around 35 years for railways to make the substantial switch to diesel that the British Navy made in about 3 years just before World War I.

James Watt, the inventor of the Boulton-Watt steam engine, had imagined its application in the transportation industry, powering water-wheel ships and locomotive engines, but he had never attempted to make those fantastical inventions a reality. That work was left to other inventors, churned up in the froth of Watt's wake, eager to capitalize on the future he had made possible. Credit for the invention of the steam locomotive goes to a man named Richard Trevithick, born in Cornwall, England in 1771, 50 years after Thomas Newcomen invented his steam engine in the same region. Physically strong and over six feet, two inches tall, Trevithick became known as the Cornish giant. At a young age he went to work for his father at a mine and became fascinated by engineering. Improving upon the steam engine was, of course, the great problem of the day, and Trevithick was like other young men in giving it a try.

As an engineer at a mine, Trevithick developed a smaller, higher pressure steam engine that came into demand at other mines for its usefulness in hauling up ore. But his creative genius honed in on the problem of how to produce steam-powered loco-motion. With his lighter weight engine he invented a miniature locomotive in 1796, and then built a larger road locomotive within five years, offering seven friends a lift on Christmas Eve, 1801. Still, the steam quickly ran out and the problem of how to power a locomotive for longer journeys remained to be solved.

Like James Watt before him, Trevithick needed capital invest-ment to support his work. He traveled to London to find backers among scientists and financiers. Despite his own earlier enthusiasm for steam locomotion, Watt actually criticized such experiments as dangerous because of the potential for explosions. Nevertheless, Trevithick secured corporate investment and built his first proto-type in 1803. The invention failed. Then a man named Samuel Homfray, who owned the Penydarren Ironworks in Wales, backed Trevithick to build a locomotive that ran on rails, imagining that it could haul his iron ore more cheaply and effectively. The Penny-darren, as the locomotive was named, made a nine mile journey and reached speeds of nearly five miles an hour. But at seven tons, it was still too heavy for the cast iron rails and broke them on each of its three trips. Homfray felt there was no practical near-term future in the investment and removed his support. Trevithick looked elsewhere for backing to perfect his invention. He never succeeded. Like others who go unheralded in the history of energy, Trevithick ended his life penniless and mostly forgotten in his day, although George Stephenson, the British inventor who finally solved the problems of steam locomotion on rails, insisted that Trevithick be remembered for his achievements as the true pioneer. Once Stephenson and others like George Pullman, famous for his Pullman cars, made steam locomotion commercially viable, the industry erupted.

The American railway was born about 20 years after Trevithick's Pennydarren. As Trevithick had come to realize, the problems of perfecting steam locomotion were not insignificant. As late as the early 1900s, steam locomotive makers were still innovating exten-sively, making increasingly powerful, speedy, and more efficient engines.

It was around this time that the British Navy converted from coal to diesel. With no military prowess at stake, there seemed to be no similar urgency or impetus in the rail industry to convert as well. In 1918, the first commercial diesel-electric locomotive was built by General Electric for a street car line in New York City, made possible by the invention of switchers. Thereafter, the diesel locomotive began to appear on national rail lines, but its widespread adoption was slow.

In 1930, General Motors bought the Electro-Motive Corporation, the premier maker of diesel locomotives at the time. GM also bought the Winton Engine company, Electro-Motive's chief supplier of diesel engines. By 1939, the first mass-produced diesel locomotives came on the market. Winton and Electro-Motive were formally merged within General Motors in 1941 and renamed the Electro-Motive Division (EMD). EMD was a market leader and in some years had almost 90 percent market share of new diesel locomotives. (In 2005, after 75 years, GM finally sold EMD to a group of investors, bringing to an end GM's long affair with America's railroads.)

By the 1940s, after 125 years of innovation, development, and implementation, steam locomotives were a mature business with most if not all of the bugs worked out of the system. Even so, the number of diesel-powered locomotives kept growing, and diesel fuel took over more and more of coal's market share. What made the diesel engine a compelling alternative in the locomotive industry? From an engineering standpoint, the principal advantage was in its thermal efficiency. In simple terms, a diesel engine is much better at converting the energy resident in diesel fuel into locomotive power than a steam engine is in converting coal. The best steam engines in the nineteenth century were only able to convert six percent of the energy in coal into locomotive power. The rest of the energy, 94 percent, was mostly blown out the smoke stack as unused heat. A burst of innovation in the early 1900s elevated the efficiency up to between 10 and 12 percent, but essentially 90 percent of the energy in every shovelful of coal was wasted.

Contrasting steam engines to diesel we see a remarkable difference. By the middle of the twentieth century, diesel engines were operating at an efficiency of 30 to 35 percent. Today, some diesel engines can achieve in the mid-40 percent efficiency range, the effective limit of their efficiency due to the laws of physics. In other words, a diesel engine today will always throw away about 60 percent of the energy in a gallon of diesel fuel, and only 40 percent is directed toward the useful purpose of turning gears and wheels.

As naval strategists had discovered earlier, diesel locomotives were more efficient in terms of work than their coal-fueled counterparts. The compelling economics were further enhanced by the price of diesel itself. At the time diesel power became an option, the United States still had abundant and growing supplies of crude oil. Diesel fuel, at about eight cents per gallon, was quite inexpensive. Moreover, because more energy was "packed" in a gallon of diesel fuel than an equivalent volume of coal, the range of a diesel locomotive was substantially greater. This advantage in "energy density" combined with superior energy conversion meant that a diesel engine could go over 500 miles without refueling, whereas a coal-fired locomotive could typically only go 100 miles. Time and cost savings were substantial and there was less need for fueling infrastructure along the tracks.

There were other, somewhat less important, but also compelling reasons for the switch. Steam locomotives needed a substantial amount of water, which had to be supplied along the tracks in water towers; diesel locomotives didn't. Diesel engines required less maintenance and operating labor; they had more traction power, cleaner exhaust, and lighter axle loads that saved track and bridge maintenance, too. In short, diesel was an absolutely compelling substitute over coal in the railroad business. In fact, as more rail companies converted, the competitive advantage created by such a switch forced rivals to follow suit. If they didn't, their businesses were so disadvantaged that they faced bankruptcy. In the commercial world, the threat of bankruptcy is as strong an incentive as any military threat encountered by Churchill in his decision to revamp the British Navy. The fact that this conversion took decades speaks less to the advantages offered by diesel over coal than to the fact that both supply chains were not experiencing the kind of extreme pressure that led Churchill to make his fateful decision to convert his navy.

The growth in cumulative diesel locomotive sales between 1931 and 1975 is illustrated in Figure 3.2. As you can see, the steepest growth period was in the 1950s and 1960s. Although the railways were only one more aspect of the overall growing U.S. economy, they represented a significant new market for petroleum, coinciding with the steepest period in oil consumption growth.

In general, oil was as compelling commercially as it was militarily. Although oil had been making notable inroads in the economy since World War I, and had been prophesied by Reid Sayers McBeth as the superfuel of the future, it wasn't until the 1950s and 1960s that oil products truly accelerated their reach into the commercial realm. Cars got bigger. People moved from urban centers to larger houses in the suburbs, and used those bigger cars to commute.

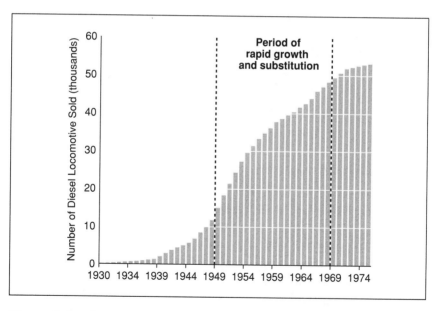

Figure 3.2   Cumulative U.S. Diesel Locomotive Sales: 1930-1975
(*Source: Adapted from data in* Diesel Locomotives: The First 50 Years)

Oil became a predominant fuel in heating those larger houses, too. Oil-fired electrical power plants were built. The airplane industry flourished as cheap air travel gave people a great sense of freedom and ease. It was an age of wealth and prosperity, fueled by cheap oil. The demand for that oil grew like never before. People didn't think twice about where all this energy was coming from. The sense that cheap, plentiful energy was an American birthright had never been stronger.

From 1950 to 1973, the world economy grew at an average of 4.9 percent. In particular, from 1961 to 1969, growth was so strong that many economists, politicians, and business leaders began to talk about a "new economy" in which the old rules of economic ups and downs no longer applied. (Previously, this term had been used in the 1920s, and we heard it again during the high-tech boom of the late 1990s.) In the 1960s, this strong economic growth catalyzed supernormal demand for oil of nine percent per year. The transition from steam trains to diesel had put the final nail into the coffin of coal-powered transportation. We had become firmly and resolutely addicted to cheap oil, and it would only take a little group called OPEC to help bring another energy break point to the rapidly industrializing world.

He Who Controls the Oil Controls the World, Part II

In the early 1900s, oil was a fuel almost completely controlled by a few companies—particularly Standard Oil and Shell, with a number of other lesser rivals in different regions of the world. Those giants exploited producing regions and set prices in the marketplace without concern for any power but each other. When oil became a strategic military commodity, and cheap oil was found to be concentrated in particular regions like the Middle East, the multinational oil companies had a new dynamic to reckon with in terms of the geopolitical maneuverings of the world powers. As Britain, France, Germany, and America jockeyed or fought for control of the oil-producing regions, the multinational oil companies served as instruments of government policy, even as they also seemed to serve the needs of oil itself. The needs of oil were simple: it demanded a price that was high enough to be worth exploiting, yet not so expensive as to be disruptive of the marketplace. In order to maintain that equilibrium of cost, price, and profit, oil companies frequently needed to cooperate among each other—and an extensive and almost bewildering web of alliances grew. But although the oil companies needed to negotiate with producing nations for concessions, once they got a foot in the door, those companies had incredible leverage in terms of expertise and capital, allowing them to be firmly in control of their host nation's oil fields.

The Red Line Agreement was one example of this expression of corporate power. In conjunction with the governments of Britain and France, the oil companies divided up the Middle East like a pirate's treasure. Later, when the American oil companies (with the backing of the U.S. State Department) were able to force their way into this cozy club, the primary rivals met for a weekend of grouse hunting in Achnacarry Castle in northern England to hammer out the new terms in the so-called "As Is" Agreement of 1928. This deal only confirmed the natural order of things, albeit while letting a few extra players into the game, leaving us with seven giant multinational oil companies, the so-called seven sisters: Standard Oil of New Jersey (later to become Exxon), Royal Dutch/Shell, British Anglo-Persian Oil Company (later to become BP), Standard Oil of New York (later to become Mobil), Texaco, Standard Oil of California (later to become Chevron), and Gulf Oil. In the past 20 years the seven sisters have consolidated into four. Exxon and Mobil merged to become ExxonMobil. Chevron acquired Gulf and Texaco and is now just known as Chevron. Shell and BP are still whole, though BP acquired Amoco, which was Standard Oil of Indiana.

There had always been suspicion cast over the motives and loyalties of the seven sisters. Rockefeller's Standard Oil was so feared that it had been broken up by antitrust regulation. Churchill led his nation into buying 50 percent of Anglo-Persian, later to become

Anglo-Iranian, in order to ensure its commitment to the British well-being. If there was resentment and doubt in consuming countries, these feelings were strong in producing countries also. After all, the multinationals focused on maximizing their profits upstream, at the point of production, in order to lessen the taxes they needed to pay downstream in the end-market. Too often the producing countries felt exploited over revenues and overlooked in the decision-making process.

When those producing countries had the power to do something about that exploitation, they did—sometimes successfully. For example, in Russia, after the Bolshevik Revolution, the new government nationalized the concessions that had been held by the foreign corporations, and in 1938, Venezuela demanded better terms for its oil contracts, threatening to also nationalize its concessions. In the case of Venezuela, a new agreement was eventually reached in which the Venezuelans received higher revenues, while the oil companies still recorded strong profits. When a more radical government came to power in 1945, its oil minister, Perez Alfonso, would demand a 50-50 share in all profits with foreign oil companies. Alfonso would later become one of the key founders of OPEC, and his idea for a 50-50 split with foreign companies followed him to the Middle East, where this profit sharing was not only a new precedent, but a call for action against oil company hegemony.

Saudi Arabia was next to demand a 50-50 split. Conscious of wanting to preserve its special relationship, but not wanting to set a precedent of a 50-50 arrangement, Aramco and the U.S. State Department worked out a compromise in which fewer taxes were paid by Aramco to the U.S. government, freeing up money that could be handed over to the government of Saudi Arabia in *de facto* foreign aid. In consequence, since the producing countries were now becoming partners in profits, they began to insist that oil be sold at a regulated price, and that those prices be made public. This fixing of prices became a feature of oil markets as OPEC rose to power beginning in 1960.

By 1951, the new leader of Iran, Dr. Mohammed Mossadegh, sought a similar 50-50 agreement with Anglo-Iranian (BP). The

British government considered invading in order to secure their oil trust, but realized that producing oil in an occupied country would be difficult, and such intervention would not be approved of by the United States or the rest of the world. Negotiations fell apart, and the Iranians took over the oil production and refinery facilities, forcing British corporate employees and diplomats to leave the country. Very quickly, Iran's oil business came to a complete standstill because no one left in the country had the expertise to keep BP's operations running. Instead oil reserves in Saudi Arabia were tapped by the multinationals to make up the gap in oil lost when Iranian production was cut off. Lacking the capital and expertise to exploit its own oil reserves, the government of Mossadegh fell in 1953, and a military coup restored the Shah, who entered into new negotiations with the foreign oil companies. A new arrangement was struck in which Iran would no longer be solely reliant on British Petroleum, although BP retained a 40 percent share in the new agreement. This episode of assertive nationalism, although ineffective in its ultimate aims, did have a strong impact on world oil supplies, creating a shortage during the height of the Korean Conflict that impeded war operations.

In 1956, the Suez Canal Crisis provided another moment of doubt and uncertainty that influenced the power balance between the multinational companies and the producing countries. When Egypt took control of the canal, and the Syrians sabotaged the Iraq Petroleum Company pipeline to the Mediterranean, the double blow was a great threat to the British sense of security of supply. Britain and France joined together to retake the canal, only to have the international community go against them. Their actions weakened the position of British oil companies in the Middle East with respect to the American companies and is one of the factors that has lit the kindling of Arab nationalism in the half century since.

The next threat to the stability of the world oil market came from smaller, independent oil companies. Tycoons like J. Paul Getty in Saudi Arabia, Dr. Armand Hammer in Libya, and Enrico Mattei in Iran, broke ranks with the multinationals to make deals with producing nations that undercut the arrangements already in place for decades. For the leaders of those countries, this reinforced the fact that significantly increased revenues were possible with better deals. Smaller, independent oil companies could help their nations capture more value for their oil. Unfortunately for those producing nations, there was an excess in global oil supply in 1957, and their newfound power could only be leveraged when demand outpaced supply.

In the United States, supply was being restricted through a long-standing combination of government regulation on domestic production called "prorationing" and voluntary restrictions on imports by the multinationals. The goal in the 1950s was to keep U.S. prices high enough to protect the economic livelihood of domestic oil companies. This protectionist stance also served to ensure security of supply, which emerged as a hot issue during the Suez crisis. Nevertheless, higher-cost U.S. oil could not come close to competing with the compelling low cost of the prolific Middle Eastern oilfields. Global market forces were too strong. Voluntary import restrictions were a porous and crumbling barrier against the cheap foreign oil that kept pouring into the United States. By 1958 almost 40 percent of domestic production was shut-in to combat the price-eroding effect of cheap imports. The dynamic played out until 1959, when President Eisenhower imposed the Mandatory Oil Import Control Program (MOIP), an import quota system to protect the livelihood of the domestic oil industry.

MOIP did its job throughout the 1960s. Eisenhower's legislation protected the domestic industry from cheap imports, saved oil workers their jobs, and preserved the nation's ability to produce the strategic commodity. With imports restricted, utilization of productive capacity in American oilfields rose to 100 percent. But the MOIP legislation also served to create an oil glut on the rest of world's oil market, depressing international prices. By 1970, U.S. prices were $3.18 per barrel, compared to $1.30 per barrel elsewhere. It was this divergence in oil prices, and the cartel-like role of the seven sisters in juggling the world's barrels and dictating prices, that sowed the seeds of discontent among big producers like Venezuela, Libya, and the giants of the Middle East.

After all, from their perspective, serious money was being left on the table.

OPEC was officially formed in 1960, in Baghdad, in part to better deal with these twin disadvantages of strong multinational companies and U.S. protectionism, which kept the producing nations' position strategically weak.[3] This was a significant concern for countries who were collectively among the poorest on Earth.

In the poker showdown between OPEC, the major multi-nationals, and the oil-consuming countries of the United States and Europe, Libya emerged as the wild card. Libyan oil had only been discovered in the 1950s, but it was plentiful, high quality, and devoid of sulfur. Moreover, it sat very close to its prime market, directly across the Mediterranean from Europe. From the beginning, Libya had played the smaller oil company Occidental off against the majors to obtain better terms. When Colonel Muammar Al Qadhafi took over leadership of Libya in a coup in 1969, he brought with him a radical ideology and saw oil as his best weapon in that fight. The Libyans broke the united front of the multi-nationals by playing the companies off against one another, even as their success in obtaining better terms for their oil put pressure on the other OPEC nations to follow suit. All of this coincided with a global oil shortage caused by rabid demand growing at nine percent per year, adding critical pressure on the oil supply chain.

Qadhafi's torch-bearing actions are an important footnote. Tipping the balance in favor of the producers, he went beyond the 50-50 arrangements in 1971 and demanded higher prices and 58 percent of the take. Today, Qadhafi is still in power and Libya remains a precedent setter in concession deals. Though the deals are more complicated now, government takes exceeding 85 percent are common in Libya and other countries where the odd giant oil field may still be found. For those who want access to light sweet crude oil, it's not getting any easier and it's getting a lot more expensive.

Concerned about the rise in OPEC power in the face of sky-rocketing consumption, 23 oil companies—multinationals and smaller independents—met in New York in 1970 to formulate a common negotiating position with OPEC. In the past, the multi-nationals had more or less informed the oil-producing nations what the price of oil would be; now, there was a sense that the power balance was being irrevocably shifted. This pressure was increased by the fact that the rate of oil production in the United States had finally reached a physical maximum, or peaked. A geologist at Shell named M. King Hubbert had predicted, a decade before, that it would happen. Suddenly, those fears were being realized as 'Hubbert's Peak' was reached. The USSR overtook the United States in oil production volume just as demand for oil was sky-rocketing and the United States was being forced to increase the amount of oil it imported. A giant oil field was found in Prudhoe Bay, Alaska, around the same time as another was discovered in Siberia, but these would be among the last two so-called elephants discovered in the world. Finally, after a century of exploration, all of the low-hanging fruit had been picked.

For its part, the U.S. government was seemingly unconcerned about the growing tension between OPEC and the independent oil companies. The multinationals tried to explain the stakes, but their warnings were ignored, perhaps because they were not truly trusted to look after interests other than their own. The multi-nationals met with OPEC representatives in Vienna in October, 1973 to determine what would be done about oil prices. This time, it was the OPEC nations who intended to dictate the terms. Through their state-owned oil companies, they planned to increase prices aggressively.

In a quirk of history, the meeting in Vienna took place just as war between Israel, Syria, and Egypt broke out. The so-called Yom Kippur War put a halt to negotiations, even as it steeled the deter-mination of OPEC. When it was learned that the Nixon Admin-istration was sending military aid to Israel, OPEC retaliated. It decided not only to raise prices as had been planned, but also to cut production by five percent per month to those countries that supported Israel. The oil embargo had begun.

After having already risen steadily from $4.00 per barrel to $10.00 over the year prior to October 1973, the embargo drove the spot price of a barrel of oil up another two-and-half times from $10.00 per barrel to $26.00[4]. In England, this coincided with a coal-miners' strike, leaving that country literally in the dark. For the first time in 80 years, kerosene lamps were put into use to light the great financial houses of London. In the United States, the embargo created a price shock that had never been seen before. Cars lined up for gasoline and customers were willing to pay any price. In fact, the hoarding mentality was so bad that many cars were idling in line with a near full tank of gas, so anxious were people to make sure that they had topped up. To make matters worse, that winter was extremely cold, putting heating fuel at an all-time premium and increasing the insecurity and anxiety that people felt about being cut off from their suddenly precious oil. Our energy birthright, so strong only months before, seemed to have collapsed like an imploded building.

The multinationals, once so powerful, were caught in an impossible situation. They still needed to distribute the (reduced) supplies of oil from the OPEC nations to their customers. To divert supplies from other sources in order to make up for the embargo deficit would undermine OPEC's intentions and risk their wrath. But to not supply a country when global oil was available would ensure the anger of that consuming country. The multinationals tried to conduct themselves using the principle of "equal misery," sharing the pain among all according to the wishes of OPEC. But each country demanded that they be considered a special case. The Netherlands, for example, had been singled out as a European supporter of Israel, but forced Royal Dutch/Shell to meet its needs. To the British, worried about the Dutch influence of British Shell since its earliest days, this oil was diverted from their own needs, and confirmed an ancient prejudice that Royal Dutch could not be trusted because of its "foreign" influence.

The embargo was over in Europe by the end of 1973, and over in the United States by March, 1974. Nevertheless, the psychological damage was done, and OPEC had our full attention. For a generation, people did not feel secure about oil again. What's more, even though the embargo ended, oil prices stayed high for much longer, contributing to an unhealthy economy throughout the 1970s. Instead of pointing their anger at the Middle East and OPEC, most Americans blamed the oil companies for the misery, questioning their loyalty and motives like never before. It didn't help that those oil companies showed record profits during those years because of the high price of oil.

In response to the crisis that was shaking his nation, President Nixon called for a new Manhattan Project that would lead to energy self-sufficiency in the United States by 1980—a response that sounds familiar today as politicians 30 years later react to the energy pressure we are facing now.

Pressure Buildup, Break Point, Rebalance

Global tension, anger at oil companies, drastic government action, frustrated consumers, steep and volatile prices, economic uncertainty, harsh conservation measures, no relief in sight. From the OPEC Embargo in 1973 until the early-1980s, the pressure surrounding energy was unrelenting.

When pressure builds in the energy cycle, it's analogous to the way steam builds up in a steam engine. The central part of a steam engine is the boiler. It's in the boiler that the energy in a pound of coal is turned into pressurized steam energy, which is then converted and transferred to the big wheels through a system of pistons, cams, gears, and levers. It's the train engineer's job to make sure that pressure in the boiler doesn't build up beyond a certain point, lest the big steel tank catastrophically blows up. In the early days of steam engines, a boiler blowing up was not uncommon. Later, as knowledge and control systems improved, overpressuring became less of a danger. Safety relief valves "blew off steam" if the pressure rose beyond a certain point and helped to avert the danger of explosion.

Such is the case in the energy cycle. Pressure builds to a point where the relief valve starts blowing before a break point is reached. Rebalancing, or "letting off steam", is necessary to bring the system back into equilibrium. Although a boiler depressurizes in a matter of minutes, energy cycles take several years, sometimes decades, to let off steam. In the history of energy break points, we have always been able to avert catastrophe, but that doesn't mean the temporary pain and uncertainty has been insignificant to the people living through such an era.

How do we know when an energy break point is approaching? An engineer watching a steam engine knows that pressure is building when he sees the pressure gauge rise rapidly. Soon, the safety valve blows with a deafening sound of steam rushing out of a pipe like a tea kettle whistling when it reaches a boil. In the energy cycle, the main pressure gauge shows price. We watch (and start to sweat) as the prices of oil and its derivative petroleum products rise. We worry about what it will do to the economy, and how it will impact our lifestyles. Increasingly drastic attempts are made to ease the pressure by opening up a valve that brings on more energy supply. (Think about how often OPEC is asked to open up their pipes whenever prices rise, or the U.S. president is pressured to tap the Strategic Petroleum Reserve.) Ordinary people wait to see what will happen, hoping that the boiler will not overheat and the economy will not go into recession. But as the price gauge keeps rising, energy-intense industries, especially inefficient ones, start closing their doors because they're losing money due to high energy prices. Employees are thrown out of work and the pipes and bolts of the economy strain under the pressure. In the worst case, some trigger, like a geopolitical event, a hurricane, or other disruption in supply, sends the needle of the gauge into the red zone. Shortages, pain to industries and the financial markets, dire economic conditions, even wars can erupt over energy supplies—the global equivalent of a boiler blowing up. It's difficult to predict how that will happen without knowing all the circumstances and ramifications. In a boiler, the pressure gauge measures that pressure. It doesn't tell the engineer about other things that may be going wrong. To monitor those other concerns, the engineer has other gauges. In the same way, price is what tells us that something is amiss with supply and demand. But other things could be wrong in the energy supply chain too, other issues that could lead to a break point.

The analogy with the steam boiler depicts what happens to the energy evolution cycle metaphorically, but to understand the actual dynamics of energy, past or present, we need to think about energy in terms of complex systems of supply chains. It's not sufficient to think about oil by itself, coal by itself, or natural gas by itself. Watching one pressure gauge for one fuel may tell us that something is wrong, but it doesn't tell us much about the whole system of supply chains that are contributing to all the useful work that is being performed in the economy. Regardless of whether that energy is going toward toasting bread or pulling trains, we need to consider the simple, yet important concept of the energy mix.

A nation's energy mix is the quantity of each primary energy source that it calls upon to meet its economic needs. Figure 3.3 shows two pie charts. The one on the top represents the current energy mix for the United States; for contrast, the one on the bottom is France.

Note the stark difference in the mix of fuels used to power the day-to-day activities of these two industrialized countries. France has built a large nuclear power base, so fossil fuels—oil, natural gas, and coal—only make up 57 percent of its mix. The United States, on the other hand, is 90 percent reliant on fossil fuels. From a straight energy perspective there is no right and wrong mix; however, a country's vulnerability to pressure buildup in one or more supply chains is clearly affected by composition.

The energy contained in each primary fuel in a mix works its way through a complex network of supply chains, ultimately ending up doing the work that we all take for granted. The energy to light up the lightbulbs in your house may originate from coal, natural gas, uranium, or a hydro dam. Gasoline for your car may originate from an oilfield in Texas, an oilfield in Canada or from the deserts of Saudi Arabia. The mystery of where it all comes from is one of the marvels of today's complex mix of energy sources and supply chains.

In many cases the primary fuels within an energy mix can act as substitutes for one another, depending upon what type of hardware is installed down the supply chains. For example, there are all sorts of different types of hardware that can generate electricity:

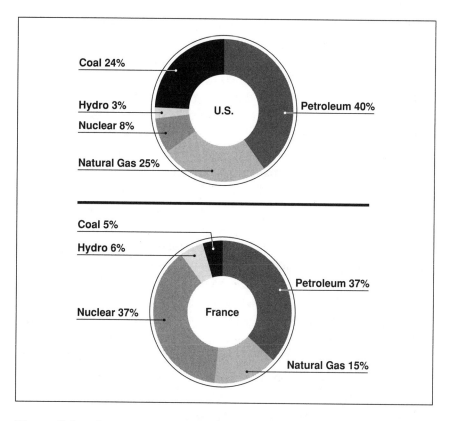

Figure 3.3   Comparison of Two Nations' Fuel Mixes: United States and France, 2004 (*Source: Adapted from BP Statistical Review 2005*)

a diesel-fired turbine; a nuclear power plant; a hydroelectric dam; a wind turbine or a solar panel, among others. Most countries have all these devices pushing electrical energy through those unsightly, big power lines that you see on major thoroughfares. To power your lights and appliances, all you need is electricity. That's all you care about. But behind the scenes the primary fuels actually compete for market share, because they're all generating the same product—electricity.

Although energy sources can often coexist seamlessly, at the end of some energy supply chains there is very little room for substitution. The vehicle you drive is a good example. It needs gasoline, which comes from oil. That's pretty much it. You can't shovel coal into your fuel tank any more than you can put in a uranium fuel rod or strap on a windmill. Other fuels like diesel can be used in vehicles, but even then you need a different engine, not to mention the fact that, ultimately, diesel still comes from crude oil. Ethanol is a fuel that can be manufactured from corn, which blended with gasoline can then be burned in a modified internal combustion engine, but it is an immature supply chain at the moment because it has historically been unable to compete with pure gasoline on cost or scale.

When do we know that things have reached a break point for a fuel? In my definition, a break point occurs when a primary fuel or an associated supply chain becomes substantially disadvantaged relative to other energy supply sources in a nation's energy mix, or relative to the emergence of a completely new supply chain. Upon reaching a break point, governments, industries, and individuals take proactive measures to mitigate the imbalance caused by the break point, and rebalancing ensues.

That explanation sounds academic, but think of it this way: your body needs vitamin C. You get your daily dose by eating oranges, apples, and peaches. Let's say the price of oranges started rising quickly due to a sudden frost in Florida. Oranges become a substantially disadvantaged source of vitamin C. After the price of oranges rise above your threshold price point, you will probably buy more apples to compensate for your vitamin needs, or substitute peaches even though you may need to eat a great number of apples or peaches to meet your vitamin C needs. If this happens, we can say that oranges have reached a break point where it was necessary to take a different course of action to continue to afford the vitamin C you need.

The term "disadvantaged" has further meaning. The first thing that comes to mind when we think about a fuel being disadvantaged is price. It becomes too expensive. Disadvantaged actually encompasses a broad set of possibilities, however. A fuel becomes "disadvantaged" when:

- It becomes too expensive relative to substitutes—Price reaches a point where companies and individuals start actively seeking alternative ways of producing the same end work;

- Its utility to consumers becomes compromised—A classic example of this is when society realizes that a fuel has become too dirty to continue using. The fact that nobody wants to build more nuclear power plants in the United States is largely due to storage concerns for the by-product radioactive waste and the fear of disasters like Chernobyl and Three Mile Island; therefore, despite other attractive aspects of using uranium, its utility as a fuel is severely compromised, or "disadvantaged" in the eyes of the public;
- Its secure supply can no longer be guaranteed—If a fuel can't be available when people want to use it, then it is not of much use, especially if there are alternative ways of doing the work. Society will often pay huge premiums for security of supply. Wars in the twentieth century demonstrated that nations that guard the interests of their society's energy addictions are prepared to use military force;
- It becomes a strategic military liability as with the example of the 33 percent advantage of oil over coal. The military is the least likely institution to compromise on a disadvantaged fuel.

Each nation's citizens, corporations, and governments react to a break point in different ways, because each nation's energy mix provides different opportunities for substitution, and each nation has a capacity to assert influence over its populous, or conduct war to secure more supply. No one is afraid of Luxembourg going to war in the Middle East if its crude oil supply is too tight, but it's not inconceivable for a nation like China to mandate that every urban vehicle be fitted with diesel or hybrid engines.

The 1970s Break Point

Let's look at the 1970s break point now to see how the United States, in particular, emerged from the pressure buildup.

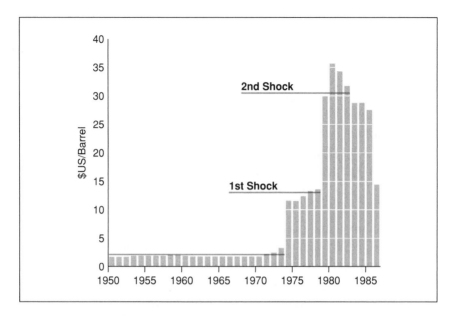

Figure 3.4   Sticky Pressure Gauge: Nominal Crude Oil Prices, 1950-1986 (*Source: Adapted from BP Statistical Review 2005*)

Although price is often the primary warning sign that a fuel is becoming disadvantaged, it did not provide much advance warning for the crude oil situation in the early 1970s. Figure 3.4 shows nominal prices between 1950 and 1986.[5]

Note that oil prices were level in the years preceding both the 1973 price spike and the one in 1979. A lot of that was because most of the world's oil was traded on prearranged, fixed-price contracts as opposed to free-market spot prices like today. In effect, the pressure gauge measuring price was very "sticky" and unable to measure the pressure building up from growing world demand and OPEC's geopolitical posturing. In effect, price was an ineffective pressure gauge.

Executives at the multinational oil firms were aware of the tenuous situation, yet their voices went unheard even as the Yom Kippur War began in 1973. Then in 1979 the pressure became worse. The Shah of Iran, a friend of the West and custodian of 5.5 million barrels per day of production[6] (9.1 percent of world oil supply at the time), was deposed in the Iranian Revolution in early 1979. Being an especially close friend of the United States was part of the problem. Not only were the Shah's cultural values at odds with Islamic fundamentalists, but he represented remnants of oil colonialism incumbent since the days when the British ruled the reservoirs with the Anglo-Persian Oil Corporation. It did not seem to matter that the oil colonialists were now viewed to be the Americans, rather than the British.

Across the border from Iran, in July, 1979, a 42-year-old political pit bull named Saddam Hussein took the reins of power in Iraq, producer of 3.5 million barrels per day of production. Next, on the eastern border of Iraq, the cold war became hot again as the Soviet Union invaded Afghanistan on Christmas Eve, 1979. Though Afghanistan was a country with no oil, the invasion would have far reaching consequences in terms of radicalizing certain elements of the Arab-Muslim population against the West, ultimately setting the stage for the geopolitical pressure on today's oil supply.

Finally, to cap it all off in September, 1980, Saddam Hussein's army attacked Iran over long-standing rivalries that had pitted Mesopotamians against Persians for centuries. Aside from the immeasurable human tragedy resulting from the bloody eight-year Iran-Iraq war, 5.6 million barrels per day of oil was taken off the world's market in three short years.

As an important side note, when oil-producing countries undergo radical political upheaval, their oil production is drastically impaired for a long time, if not permanently. Figure 3.5 shows three oil production graphs, one each from Iran, Iraq, and Russia.

For Iran and Iraq the major upheaval was the eight-year war starting in 1979. For Iraq the ups and downs continued with the Gulf War in 1991 and the more recent U.S.-led invasion in 2003. Neither Iran nor Iraq have restored their pre-1979 production volumes. For Russia it was the implosion of the Soviet Union in 1989. Over a six-year period, Russian oil production dropped by almost half to six million barrels per day. Though Russian production has been on a rebound since 1999, it has yet to achieve

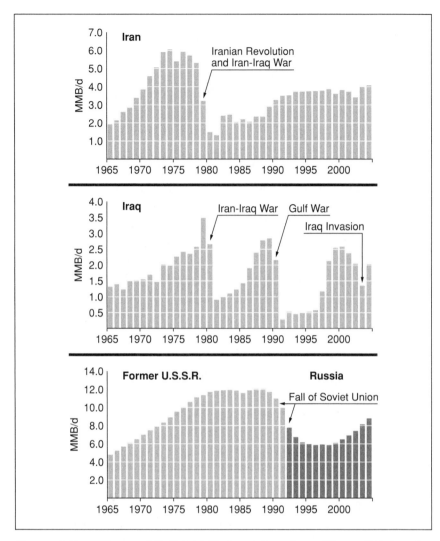

Figure 3.5   Effects of Political Upheaval on Long-Term Oil Supply: Historic Crude Oil Production for Iran, Iraq and Russia (*Source: Adapted from BP Statistical Review 2005 and U.S. Energy Information Agency*)

peak Soviet-era level of nearly 12 million barrels per day. Nigeria, Venezuela, and Indonesia are other countries where political strife has affected oil production in an on-again, off-again fashion. The real lesson is that political forces—internal or external—that get out of control tend to clamp our oil supply chains for a long time.

It's something to be mindful of when we recognize how concentrated the world's oil dependency has become on a handful of geopolitically vulnerable countries.

The 1973 oil embargo, followed by the overthrow of the Shah of Iran and the Iran-Iraq war in 1979, were two closely spaced pressure-and-break point cycles in the global energy supply chain. The uniqueness of this cycle pair was that pressure built up extremely quickly, because many parts of the oil supply chain were prepressurized with geopolitical tensions and aggressive demand growth. Although many signals of this overpressured system were present, nobody was directly watching the other "gauges" either preceding or occurring during a break point and rebalancing episode that lasted 13 years between 1973 and 1986.

Even when I talk to industry experts about the 1970s break point, the basic perception is that oil prices skyrocketed; the world economy came to a grinding halt; oil demand regressed; more oil was found, and the problem was solved. In fact, there was much more at play. The break point triggered major rebalancing efforts in every industrialized country. The world emerged in 1986 looking far different in terms of energy use than when it entered in 1973. In many ways it emerged far better.

Rebalancing the 1970s Break Point

Every growth-pressure-break-point-rebalancing episode in the history of energy has its own special characteristics. Sometimes the pressure cycle builds gradually and comes to a head, but the transition to rebalancing is achieved through fortuitous circumstances, as happened when the demise of whales and whale oil was offset by the timely emergence of rock oil. Other times, pressure builds very quickly and it takes much effort from industry, government, and society to rebalance. Such was the case of the oil shocks of the 1970s.

The world's economy slowed down dramatically in 1974 and 1975 (down to an average 2.3 percent GDP growth from a blistering 6.8 percent in 1973), and again for a few years following

1979. By 1985, OPEC was ratcheting up production to 18 million barrels per day, bringing enough supply on to make the markets feel that the price shocks were over. Non-OPEC supply from Alaska and the North Sea were important factors in alleviating pressure, too. But these were not the core reasons that the break point of the 1970s came to an end. Oil had been disadvantaged relative to other energy sources, and economic growth was threatened. Action was needed on the demand side of the equation, as much as on the supply side. The factors that really made a difference in the 1970s break point era were:

- The implementation of government policies in many industrializing countries that forced utilities, businesses, and individuals to conserve oil and buy appliances, including cars, with better energy efficiency.
- The implementation of government policies that forced manufacturers of vehicles to improve fuel economy.
- A massive buildup in coal and nuclear power plants that squeezed oil out of the electrical-power-generating market.
- A large global buildup in liquefied natural gas infrastructure, including tankers, that helped countries—in particular, Japan—to become less reliant on oil.

The impact of these actions was staggering in magnitude, for collectively they helped arrest the year-over-year demand growth for crude oil, which was compounding by nine percent per year prior to 1974, down to one-and-a-half percent per year after 1985 (see Figure 3.6).

In the United States, the break point and subsequent rebalancing were striking. Figure 3.7 shows the progression of the U.S. energy mix since 1965, revealing the proportions of primary fuels that go into all end-use markets, from transportation to electrical power. At the very bottom is a thin slice that represents hydroelectric power. It hasn't been growing in the last half century because all the major rivers were dammed up by the 1950s. On top of hydroelectric is oil, which you will note is the highest volume primary fuel source. Coal, nuclear power, and natural gas are layered

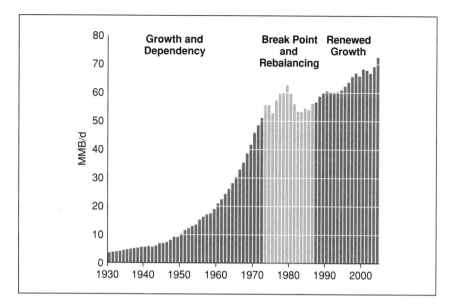

Figure 3.6   World Crude Oil Demand, 1930-2004: Full Cycle Energy Evolution (*Source: Various and ARC Financial*)

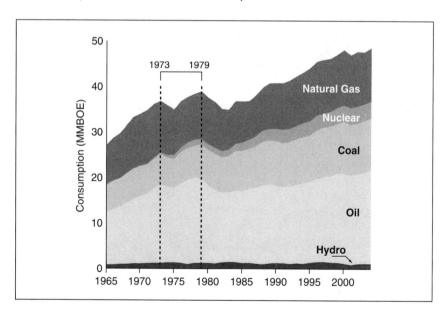

Figure 3.7   Evolution of the U.S. Energy Mix: All Primary Energy Sources Converted to BOE (*Source: Adapted from BP Statistical Review 2005 and ARC Financial*)

on top successively. For now, most of our interest lies between the two vertical dashed lines that mark 1973 and 1979. First note how the demand for all the energy commodities slope downward immediately after each of the vertical lines. That's the effect of the slowing economy as a consequence of the price shocks, the place we reached an energy break point. As I've discussed before, the economy and all energy supply chains are inextricably linked.

To really understand what happened in the 1973 to 1986 break point and rebalancing period, however, I need to show you an energy mix chart that just supplies electrical power. This time, I'm going to show the mix as a market share diagram, so all fuels add up to 100 percent.

In Figure 3.8 I have highlighted vertical dashed lines for 1973, the start of the break point period, and 1986, the end of the rebal-

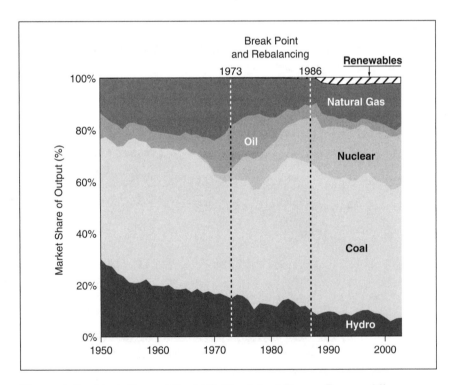

Figure 3.8 Evolution of the U.S. Electrical Power Energy Mix: Expressed as Percent Market Share, Before Converting to Electricity *(Source: Adapted from U.S. Energy Information Agency and ARC Financial*

ancing, when the oil price pressure gauge really came down hard. From a market share perspective, hydroelectric power has been losing ground since 1960. Again, that's because there were no more major rivers to be dammed up. The bulk of electrical power was, and still is, generated by coal. Even today about 50 percent of U.S. electricity comes from coal-fired power plants, not surprisingly because coal is plentiful in the United States. In 1973, 45.5 percent of the power generated came from coal. Between 1965 and 1973 the market share of oil-fired generators was increasing as the economy was growing rapidly. By 1973 oil was 17 percent of the power generation mix, and natural gas was 18.3 percent. The very thin slice at the top right are renewables like wind, geothermal, and solar power. Note that in 1973 there was very little nuclear power in the mix. That's because the technology was just emerging, and building a nuclear power plant was exceedingly expensive.

What Figure 3.8 shows very clearly is how nuclear power and coal power squeezed oil out of the power generation market. To a lesser degree they squeezed out natural gas too. By 1985 oil had fallen to 4.1 percent of the power generation market and today it sits at under 3.0 percent. That means, despite what popular wisdom tells us, conserving electricity can never wean us off our dependence on foreign oil. Getting rid of your gas guzzler, well, that's another story.

How did this big squeeze happen? There were three important drivers. First oil was economically disadvantaged as a fuel for generating electricity relative to coal. Second, nuclear power—like kerosene in the days of whale oil—was a technological savior waiting in the wings. Not only did nuclear power serve as a good substitute, it was a large-scale substitute that could be introduced in a relatively short period of time (remember 13 years is an eye blink when it comes to changes in the energy supply chain).

The third part of the story is significant and not well remembered. In 1978, Jimmy Carter's administration introduced the Fuel Use Act. In effect, utilities were legislated against using either oil or natural gas to generate electrical power. Not only was oil suddenly disadvantaged by price, but now it was also disadvantaged by legislation. It all added up to coal and nuclear power taking away market share very fast between 1978 and 1986. At the end of that time period, the rebalancing exercise was complete in the electrical power market and government policy had been a major catalyst.

But the bulk of oil was, and still is, consumed for transportation. Like in the power market, there were three main factors that helped in rebalancing the transportation segment. First, there was the effect of price. Between 1973 and 1985, gasoline prices rose from 39 cents per gallon to $1.20 per gallon, so there was a personal financial incentive to rebalance the wallet by buying a smaller, more fuel efficient car. To assist consumers in that direction, the government imposed the Corporate Average Fuel Efficiency, or CAFE, standards on the automakers in 1976. Under the legislation, the Detroit automakers were mandated to improve the dismal fuel efficiency of big, gas-guzzling vehicles from an average 12.9 miles per gallon in 1974, to 27.5 miles per gallon by 1990. As we'll see in more detail in Chapter 4, the legislation catalyzed a lot of improvement, though actual average fuel economy on the roads has stalled out at about 20 miles per gallon.

But the burden wasn't entirely on Detroit. The consumer had to pitch in to help too—by slowing down. The National Maximum Speed Limit was introduced in 1974 to reduce fuel consumption and, as an added bonus, improve safety too.

By the time 1986 came around the price of oil had dropped from an annual average 1980 high of $U.S. 35.69 per barrel, down to $U.S. 14.43. Gasoline prices had fallen back to 93 cents per gallon. The economy was growing again too. That's all most people remember. But the rebalancing that went on behind the scenes would change the United States. and many other nations of the world for the better.

Notes

1   Found in *Nukespeak*, by Hilgarten, Bell & O'Connor, which cites "Notes and Comments," *The New Yorker*, vol. XLIX, no. 42, 10 December 1973, p. 37.

2   *Oil: The New Monarch of Motion* by Reid Sayers McBeth,
    p. 2; 1919, Markets Publishing Corp., New York.
3   The founding members of OPEC were Saudi Arabia, Iran,
    Kuwait, Iraq, and Venezuela. Later membership came to
    include Qatar, Libya, Indonesia, United Arab Emirates,
    Algeria, Nigeria, Ecuador, and Gabon.
4   Rotterdam spot price; Danielsen, *The Evolution of OPEC*,
    page 172.x
5   BP Statistical Review.
6   BP Statistical Review; three-year average Iranian produc-
    tion between 1976 to 1978 inclusive.

CHAPTER 4

TO THE ENDS
OF THE EARTH

So what is happening today? Listening to the pundits, you're bound to get confused. The alarms have sounded; the prices are up. Everyone acknowledges that we're confronting energy challenges that we've never faced before. The problem is we're not all talking about the same thing. Whether you think the end of the world is near or today's concerns will go away on their own depends on what kind of expert you are listening to at the time. There are a host of experts on the supply side, and a host of experts on the demand side—and a raging debate between them. Some think that we're running out of oil; others say we've got plenty left. Some think that world demand—especially from China—is going to push the pressure needle into the danger zone; others say that all those engines firing at the same time will cool down soon and leave us idling comfortably. Throw in the voices of those who are advocating various positions on conservation, global warming, geopolitics, government policy, and the wonders or limitations of new technological advances—and you are left with a blurry picture of what is really happening now, and how that will affect your life in the next 5 to 15 years.

Without a comprehensive understanding of the various forces affecting us today, we can't understand why the pressure in our energy cycle is rising and what that means for the near future. So let's clear some things up. We're not running out of oil, but the oil we need is getting harder to find. Neither China, India, nor the

United States is going to swallow the world's resources whole, but even a global economic slowdown is not going to turn back the clock on how much oil is consumed every year. There are no magic bullets in the form of radical technological innovations to rescue us, and yet technology in some form or another will still help save the day. It all seems contradictory, confusing, and complicated, and for the most part it is. But from these basic ideas, we can begin to get a handle on what's going on and how to bring energy's big picture into focus.

Oil Prices Rise and the Alarm Sounds

Most people don't sit glued to their television screens watching the price of oil flicker. But they do drive by the pumps every day and fill up at least once a week. That's where oil prices get our attention. Whether you're a daily commuter or a retiree who spends the summer in the RV out on the open highway, you can see and feel the impact of volatile oil prices. Increasingly, you're bound to wonder what's going on. It's the same with home heating oil. Few of us think much about the oil that has been pumped into our basement furnace when we turn on the heat, but when we get a bill at the end of a cold winter month for twice what we paid last year, we start to wonder.

Whenever I'm traveling to different cities giving speeches or attending meetings, I always ask the taxi drivers what the local price of gasoline is. Despite all the variation in price in different parts of the country, I always get the same answer: "Too high." Most people don't know why it's too high, but they know it has something to do with the price of oil. If they're talkative, they might blame, in no particular order, OPEC, the war in Iraq, American dependence on foreign oil, the Big Oil Companies, taxes, environmentalists, or SUVs.

The lack of public understanding about the issues behind oil prices is nothing new. The January 9th, 1948 edition of the *New York Herald Tribune* stated: "There is no country in the world which has the body of technical doctrine regarding petroleum in all its aspects which is possessed in the United States. There is no country which is so thoroughly geared to the power supplied by petroleum. Yet, thanks to the mixture of unsupported argument, official reticence and sheer hypocrisy which befog the subject, there can be few peoples so poorly informed of the global implications of oil production and distribution as the Americans."[1]

Personally, I challenge one aspect of this statement: Ignorance about oil prices and dynamics is not limited to Americans; it's universal among the general population of the world. But the interesting thing is that not much has changed in the half century since that opinion was written, a fact that contributes significantly to today's complex energy problems. For now, let's just deal with the issue of oil prices and show how and why they have risen. It should be noted, of course, that rising oil prices means the price of almost everything else is rising too, since our entire society—nearly everything we consume—is directly or indirectly dependent on oil and its derivative petroleum products.

When we hear oil prices quoted in the news, we are actually getting the so-called "spot" price of oil. That's the price you would have to pay if you wanted it delivered today. Delivery is usually to a hub—a storage and distribution center typified by giant, white, cylindrical storage tanks. If you want oil delivered to your doorstep, you have to pay transportation charges from the hub to wherever you are in addition to the quoted price. Of course, most of us don't think about the pipelines and trucks that deliver our oil products, but it's all part of the vast multitrillion dollar infrastructure of the energy supply chain.

When CNN or MSNBC flash up the price of oil, they're actually talking about a special light sweet grade of crude oil called West Texas Intermediate (WTI). It's a fluid and desirable grade with very low sulfur content, which is why it is called "sweet". Conversely, sour grades of oil have much higher sulfur content, making them more costly to refine and environmentally unappealing. As we saw in Chapter 1, whale oil was graded in much the same way. High-quality grades were light and clean-burning; lesser grades were heavier, impure, and more costly to refine. WTI is like the high-grade spermaceti oil of today. By direct analogy, thicker, heavier grades of oil are more akin to the blubber of a right whale. Obviously, heavier, more sour grades of oil are of lesser quality and trade at lower prices in the marketplace because they require more refining and are usually more expensive to the end-user.

Looking at the trend of WTI oil prices from 1990 to today in Figure 4.1, you can see what all the excitement is about. Prices have more than tripled since 1999, and most of the price appreciation has been in the past two years.

To understand fully what these prices mean, we need to understand the other dimensions of price. Globally, there are many different sources of oil, all of differing quality. As such, there are many different "benchmarks." WTI is a very desirable, premium grade. Brent, which is a North Sea product, is also a highly desirable for its light, sweet qualities. The difference in price between two grades of oil, usually at two different hubs, is referred to as the "differential." The differential between WTI and Brent, for example, has two dominant components: the difference in transportation costs

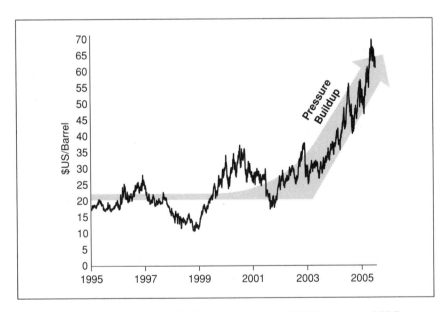

Figure 4.1   Daily Crude Oil Prices, January 1995-August 2005: West Texas Intermediate (*Source: Adapted from Bloomberg and ARC Financial*)

between two hubs, and the difference in quality. Shrewd traders watch global oil price differences very closely because abnormally wide or narrow differentials can signal a money-making opportunity. It's all part of the global electronic marketplace for oil where traders make multimillion dollar buying and selling decisions in a heartbeat without ever seeing, smelling, or touching whatever is in the pipeline or supertanker. This is in stark contrast to whale oil traders who used their well-developed senses, and shrewd business acumen, to grade and trade their products dockside in Connecticut or London.

In the marketplace you can also contract to buy or sell oil for future delivery and settlement. In this way, buyers can agree on a price today, settle up, and take delivery at that agreed-upon price next month—or 12 months out, or up to five years out and more. If you can find a seller willing to sell you oil 10 years out at an agreed-upon price, you can purchase a futures contract for that, too. Since about 1990 the market for these contracts has grown and it is now routine for suppliers and industrial consumers to buy and sell oil futures. At any time of the trading day the prices for oil futures are quoted just like the cash or spot price. That's because traders are buying and selling these contracts for future delivery and settlement. Again, in the days of whaling it was much simpler: A ship would come in with casks of whale oil. Buyers would grade the oils and offer the owner the going market rate, paid in cash on the spot. There was no futures market back then, but merchants, shipbuilders and the like would make capital investment decisions based on their view of the market price for whale oil several years out.

Futures prices are consequential for many reasons, most of which are beyond the scope of this book. In terms of understanding the pressure in the energy cycle, futures are important because they give a general sense of what buyers and sellers of oil in the market are expecting prices to be in the long term. Though the spot price of oil has risen sharply over the past three years, equally spectacular has been the rising expectation of the future price of oil. For much of the 1990s the expectation was that oil prices would revert to around $20 a barrel within two years; in other words, futures contracts were trading at $20 a barrel. By the middle of 2005, futures contracts for delivery at end of the decade had risen to over $60 a barrel.

Some argue, with justification, that futures prices are not good predictors of the actual spot price when we arrive at the future contract date. Fair enough; the marketplace is much more complicated than that, and nobody is saying that the buyers and sellers of today have a perfect crystal ball. But high futures prices are another diagnostic gauge measuring pressure in the world's oil supply chains. In this case the gauge is signaling that today's pressure build is casting a long shadow into the future.

To clarify our understanding of oil prices even further, there is also the matter of the real price of oil. Today's dollar doesn't have the same purchasing power as yesterday's dollar due to inflation. A bag of groceries that cost $10 in 1960 costs $60 dollars today. It's not that the contents of the bag has changed much; the difference in price is mostly an artifact of inflation. So, when comparing today's oil prices relative to prior years, it's often important to adjust for inflation and scale everything to today's dollars. That way we get a sense of the true relative cost between today and prior years.

The peak of oil prices in real, inflation-adjusted terms was 1980. In 2004 equivalent dollars, the high-water mark back then was $82.15 a barrel. Adjusting for inflation U.S. gasoline prices peaked in 1981 too, $2.60 per gallon as compared to $2.15 in the first half 2005[2]. As oil prices ran up in 2004 and 2005, many analysts pointed out, correctly, that we had a long way to go to get to the equivalent $80 level in 1980. Therefore, those analysts continued, we should all calm down. While there is nothing wrong with that analysis, it's a very limited notion because the absolute price of oil isn't necessarily the only issue at hand. How fast prices rise, how dependent a nation's economy is on oil, and the difference between oil price and the next best substitute, among other things, are also important concerns in understanding where we're headed.

What does all this mean to you in your car watching gas prices rise and fall at the pump? Every barrel of oil yields half a barrel

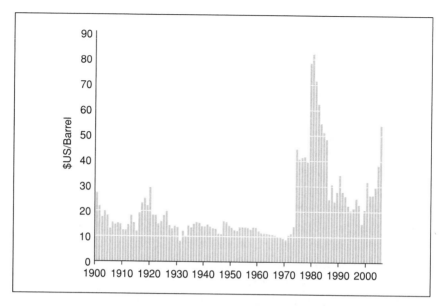

Figure 4.2   The Inflation Adjusted Price of Oil, 1900-2005: Annual Average Brent in 2004 Dollars (*Source: Adapted from Bloomberg and ARC Financial. Note: 2005 Estimates by ARC Financial*)

of gasoline after it goes through a refinery. The other half goes into other petroleum products like diesel fuel, jet fuel, heating oil, asphalt, and so on. All of these products are heavily influenced by the price of oil. Depending on the region, gasoline prices include much more than the price of the underlying barrel of oil. The retail arm incurs transportation costs to get the gasoline from the refinery to the pump. On top of that there are marketing costs to promote the product. Finally, a big slice of gasoline price is government taxes. Each country and region is different in that respect. In the United States, federal and state taxes make up about 20 percent of the price of a gallon of gas. So if the price of gas is say $2.50 per gallon, 50 cents of it goes to tax.

In fact, the U.S. federal tax on road fuels is very light compared to many other parts of the world. For instance, in Britain taxes compose 75 percent of the price of a gallon. After currency conversion, the price of a gallon of gasoline in Britain is almost three times that in the United States. This means that for drivers in the United States, the price of a gallon of gasoline is more sensitive to oil price movements because there is less of a taxation layer. On average in the United States, a $1.00-per-barrel move in the price of oil eventually translates into a 3-cent move in a gallon of gasoline.

Oil prices have been rising for five years, and when you consider the possibility of $100 per barrel for the price of oil it sounds ominous, but how does that trickle down to gasoline prices? If you run the numbers, a $100 barrel of oil implies a gasoline price between $3.50 and $4.00 per gallon in the United States. It is a lot, but it's still a far cry from the price of gasoline in heavily taxed regions of the world like Britain, France, Japan, and many others.

Oil company profits are of course embedded in price too, and are often a lightning rod of discontent among the general public when fuel prices rise. In the context of today's pressure build, caution must be exercised in recognizing what oil company profits are attributed to when prices rise quickly. So-called cheap oil, the legacy reserves that established oil companies found years if not decades ago, are admittedly highly profitable. This is akin to old inventory on the shelves that has suddenly become much more valuable. But the old stuff on the shelves is depleting and is not enough to satisfy the world's insatiable demand. New oil must be found constantly, and because the newer reserves are far more expensive to find, profitability on new barrels is nowhere near as lucrative as on the old. Indeed, oil companies must "recycle" their profits from their old, cheap barrels back into the ground so that they may find and bring to market more expensive, new barrels. As I'll discuss later on, adequate reinvestment of profits by oil companies into risky parts of the world is a key challenge and a source of today's pressure build.

If the world's oil supply chain were a hospital patient, price would be like its blood pressure. You don't have to be a doctor (or an economist) to know that something is amiss when looking at the various indicators and price charts. Like many patient illnesses, the charts can get a lot worse before they start looking better.

This Time It's Different

If you've been in this business long enough, you know that $20.00 per barrel was the rough number that analysts always felt oil prices had to average in the postbreak point period of the 1970s-1980s. It was like the normal body temperature for the industry. Many spoke of an $18-$22 range, which implied a U.S. gasoline price of about a $1.25 a gallon. If the price of oil was out of that range, it was assumed that cyclical forces would bring it back to the norm within relatively short order.

One clear, expected, and well-understood reason for deviating out of the range was seasonality. Major oil-consuming nations lie well above the equator in the northern latitudes. Naturally, the seasons induce cyclical energy demand within the course of a year. In the winter these regions need to generate heat and light, and vehicles get lower fuel economy in the cold. Not surprisingly, the first and fourth quarters of the year, the winter months, are the most demanding on the world's energy supply chains. The second quarter, which takes in spring, is the least demanding.

In 2005, for example, the difference between second-quarter demand and fourth-quarter demand was about 3.5 million barrels per day. Second-quarter demand averaged 82.5 million barrels per day; by the fourth quarter, demand was approaching 86.0 million barrels per day, or a thousand barrels a second!

Of course, knowing that these seasonal fluctuations occur, we are inclined to manage our needs during the course of the year. Just as a squirrel gathers and stores acorns for the winter, so too do we with primary fuels. Each nation, especially those in the northern latitudes, works to build up its crude oil and petroleum products inventory in time for the winter. In particular, heating oil and natural gas stores are built up during the summer so they're ready to draw down in winter. Some products like gasoline are actually more in demand in the summer. Though cars are more fuel efficient in the summer, vacationers hitting the highways put a strain on gasoline stocks during the May-to-August "driving season," something you've probably noticed by watching pump prices and listening to the news reports.

The upshot of seasonal changes is that the near-term price of oil and associated petroleum products typically reacts to the level of storage in advance of the seasons. For example, if heating oil inventories are low in September, the price of heating oil rises. At the same time, the price of crude oil rises, too, because more will have to be refined to supply the storage deficit. Conversely, if the storage tanks are full, people take comfort and prices generally fall. We should be mindful that the seasons appear to be getting more extreme due to global warming. By extension, increasingly volatile weather patterns translate directly into greater energy price volatility and a need to husband greater levels of inventory.

Prices are not only affected by the seasons, but they are also affected by vulnerabilities of worldwide supply chains, and the over-all global forces of supply and demand. One of the amazing aspects about crude oil and petroleum products is the vast supply network of pipelines and supertankers that has been established in the last 145 years. This network helps to quickly iron out anomalous price differentials around the world. For example, imagine that oil prices in the United States are too high because of shortfalls, and prices in Europe are too low due to excess. In a mere couple of weeks the imbalance can be fixed by moving oil tankers across the Atlantic, or diverting tankers from the Persian Gulf to American destinations instead of European ones. Actually, that's a bit simplistic, but in essence the world's oil infrastructure has a built-in balancing mechanism to ensure prices don't get too high or low in any one region.

One natural tendency of this network is that if producing nations start selling too much oil or too little oil into the vast supply network, it affects price in all regions. Oil is truly a global commodity. Historically, when supply or demand went askew, market forces combined with the on-again-off-again tactics of OPEC, tradi-tionally served to bring oil prices back into the normal $18 to $22 range. In short, people in and around the oil industry were condi-tioned to believe that the business was endlessly cyclical; between the seasons and the quick response of global forces, various mechanisms were always working to bring prices back into the prescribed range. In fact, those who bucked this conventional wisdom with predictions that "this time it's different" have been burned many times, further reinforcing the idea that prices can't stray out of the range for very long.

Nevertheless, this time it *is* different because of some very significant structural changes. As I previously mentioned, in much of the 1990s and early 2000s there was a reasonable relationship between price and inventory levels. If you knew where inventories were going to be, you could make a pretty good stab at price. And as also mentioned, the market generally had a belief that there were overwhelming forces at play—both on the supply and demand side—that would rectify any inventory surpluses or deficits. In that sense, the markets were like individual car drivers keeping an eye on how much gas is in the tank. If the empty light is coming on, you feel the need to fill up as soon as possible. And you'd probably do so even if the nearest gas station wasn't the cheapest. On the other hand if your gas gauge shows full, the last thing you're thinking about is pulling up to a pump.

In early 2003, however, market sentiment started to change as the world started to accelerate its rate of oil consumption, and inexpensive light sweet crude became harder to come by. Buyers of crude oil began to worry about inventory levels even if they were full. If the market could speak it was saying something like, "I don't care how full inventories are, we're going to need every drop of it to fulfill the growing demand, especially when we hit the high-consumption winter months. And I'm also worried about how we're going to fill it up again in the future!"

Two analogies help to understand the market psychology. The first is our squirrel again, feverishly storing acorns because he knows that the upcoming winter is going to be cold. Even worse, he's worried that next year's weather may be poor for growing acorns. The second analogy goes back to your gas tank. Imagine that your tank is near full, but you've heard there could be shortages of gasoline at the pump soon. It may not even be true, but because you're concerned, you're likely to be filling much more often, even at higher-priced stations. In fact, as I mentioned in Chapter 3, that's what actually happened in the 1970s. Most of the cars lined up at the long queues for gasoline at the height of the energy crisis had tanks that were better than three-quarters full.

And that's where the world is today, faced with a global hoarding mentality as a response to the tight supply and demand conditions in the vast oil supply chains. From the perspective of an evolving energy system, it's a classic pressure build. Consuming industries are looking to maintain high levels of inventory just in case shortages cause prices to go higher or even lead to interruptions. Politicians tell their constituents that everything will be fine, we'll find more oil, and technology will save the day. Yes, we'll find more oil, but no longer the cheap stuff. Yes, technology will help us, but not any time soon. These are not remedies to the acute near-term issues, not the least of which is the world's unrelenting demand for more and more oil every year. The historical $20 a barrel for light sweet crude is gone. This time it really is different; prices for oil and petroleum products like gasoline have risen dramatically and can still go a lot higher. A sustainable trend toward moderating oil prices is not forthcoming until the pressure buildup triggers the next break point and the rebalancing of our entire energy system.

The Demand Challenge

Following the 1973 break point and rebalancing period ending in 1986, global oil demand began growing at an average rate of 1.5 percent per year, phenomenally less than the nine percent per year exhibited in the late 1960s. Rebalancing had forced a new discipline of conservation and efficiency everywhere in the world. Industry shifts (like making cars lighter by replacing metal with plastic) made a tremendous difference on fuel economy. In many countries—including the United States—oil was squeezed out of the electrical power generation markets altogether by nuclear power, coal, and natural gas. In countries like the United Kingdom and Japan, a policy of high taxes on retail gasoline prices provided the catalyst for people to buy smaller cars, drive less, take public transportation, or all of the above. The tight bond between the world's expanding economy and oil consumption loosened up and we all emerged requiring less oil to transact every purchase, every mile traveled, every *everything*.

Since Reid Sayers McBeth spoke his prophetic words about oil in 1919, there has been a tight, straight-line relationship between economic activity and oil consumption. For the United States, I showed the character of that parallel relationship, between 1950 and 1970, in Chapter 3 (Figure 3.1). Now let's explore that character to understand fully how it can change after a break point.

In Figure 4.3 each year's oil consumption between 1950 and 2004 has been paired with its economic activity, or real GDP. For example, in 1950, U.S. GDP was 1.8 trillion dollars and its oil consumption averaged 6.5 million barrels per day. By 2004, GDP had grown to 10.8 trillion dollars and oil consumption to 20.7

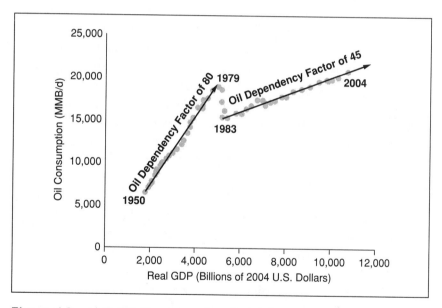

Figure 4.3   U.S. Oil Consumption Cross Plotted Against Real GDP, 1950-2004: Evolution of U.S. Oil Dependency (*Source: Adapted from Bureau of Economic Analysis, U.S. Energy Information Agency and ARC Financial*)

million barrels per day. But take a look at how the data points in the chart line up: Straight as an arrow between 1950 and 1979, and again between 1986 and 2004. The important observation is that the slope is much shallower in the second segment.

The shallower the slope, the less oil is required to lubricate economic growth. The break point and rebalancing era of the 1970s effectively cut the United States' energy intensity for oil by almost half—a laudable achievement that reduced its dependency on oil to fuel its expanding economy.

For a moment, imagine in Figure 4.3 if the data points were lining up horizontally, or flat. If that were the case, it would mean that the U.S. economy could expand without needing to boost oil consumption. All else being equal, the nation's future economic fortunes would be independent of having to seek out more and more oil supplies every year. Ideally, of course, the linear trend would be pointing downward and to the right, an enviable situation where the economy can grow, while oil consumption diminishes. If you look at the chart again, that's what was happening in the United States between 1980 and 1986, by virtue of nuclear and coal power squeezing oil out of the power markets—the rebalancing dynamic I demonstrated in Chapter 3 (Figure 3.8). Unfortunately, the dynamic came to an end after there was little oil left to squeeze out.

But that's not to say that a complete decoupling of economic growth and oil consumption cannot be achieved and sustained. Individual nations that include Japan and several in Europe have accomplished this feat in the past. Unfortunately, these countries do not represent the norm today. Because big economies like the United States, China, and a whole host of industrializing countries still have a positively correlated relationship between GDP and oil demand, the world as a whole requires an increasing amount of oil every year to facilitate economic growth.

This is a crucial point, because pressure on the world's oil supply chains will keep building so long as the global economy keeps expanding. Any global economic growth at all necessitates more and more oil every year. And from this relationship a very big myth needs to be set straight. A global economic *recession* will not cause the world to consume less oil than it already does; it will merely cause a slowdown in the rate at which our consumption is growing.

One of two things are required for oil consumption to drop below 1,000 barrels a second: a worldwide economic *contraction* where GDP actually shrinks (something that has not happened since World War II), or a break point that jolts the way energy is produced and consumed (something that has not happened since the 1970s).

So, a steep slope—reflecting a high dependency on oil to grow an economy—is less desirable than a shallow one, especially if finding new oil reserves is becoming increasingly difficult. Instead of saying "steep" or "shallow," there are different ways of indexing the slope of the line in Figure 4.3. It can also be calculated regionally, nationally, or for the world as a whole.

I call my indexed measure of the slope the "oil dependency factor."[3] A horizontal, flat slope reflects an oil dependency factor of zero; in other words, zero new oil is required to fuel economic growth. Rising slopes are positive, declining ones negative. For a sense of scale, my measure of the U.S. oil dependency factor up to 1973 was 80. After the break point and rebalancing with nuclear and coal power, it fell by half, and leveled out at about 45, though it appears to have crept up to over 50 again recently. Large, resource-based economies that are in their early stages of aggressive industrialization typically exhibit oil dependency factors of 80 and above. As Figure 4.4 shows, China and India are both averaging over 90.

Today it is the large, growing economies coupled with high oil dependency factors that are responsible for the lion's share of growth in world oil consumption. Dependency factors for nations like Japan, Britain, and France have been at or below zero since the last break point, reflecting their conscious policy efforts to mitigate oil dependency. All nations put together, the world's oil dependency factor has averaged 29 between 1995 and 2004, but notably it has been rising over the past three years and is now somewhere between 35 and 40, reflecting of course the growing influence of China in the overall average. This recent rise in the

	Oil Dependency Factor 1995–2004	2004	
		GDP[1] $US Billions	Oil Consumption MMB/d
India	94	661	2.6
China	90	1,649	6.7
Thailand	78	163	0.9
Malaysia	72	118	0.5
Taiwan	63	305	0.9
Canada	60	996	2.2
Singapore	48	107	0.7
United States	45	11,733	20.5
Korea	28	681	2.3
Australia	25	618	0.9
France	16	2,018	2.0
Japan	< 0	4,668	5.3
Germany	< 0	2,707	2.6
Russia	< 0	583	2.6
Italy	< 0	1,681	1.9
United Kingdom	< 0	2,126	1.8
World[2]	**29**	**55,655**	**82.5**

1 GDP at current prices in US billions
2 GDP based on purchasing-power-parity (PPP) valuation of country GDP

Figure 4.4   Comparison of Average Oil Dependency Factors: Various Nations (*Source: Adapted from IMF World Economic Outlook Database and BP Statistical Review 2005*)

world's oil dependency is a demand challenge that is compounding the pressure on our energy supply chains, and is leading us to a break point.

The tight, linear relationship between the economy and oil consumption means that like the seasons, the broader economic cycle influences the oil cycle. Take recent history for example. From the early 1990s to 2002, there was always some region of the world not performing well economically. For example, Asia was hit by currency devaluation, Russia had its own Ruble crisis, and the bursting high-tech bubble slowed down growth in much of the industrialized world. In essence, the world's economy as a whole was not firing on all cylinders and the threat of Y2K was looming. Therefore, global GDP growth was up and down, keeping the rise in oil consumption inconspicuous until the end of 2002. And then the pressure really started to build.

Gaining momentum into the new millennium, China began racking up supernormal GDP growth of about 10 percent per year, with urban areas seeing unprecedented growth and entire road systems being built. This economic activity trickled down to Chinese consumers, crossing a threshold of individual wealth that triggered greater consumption of energy-based products, such as home appliances and cars. This went hand-in-hand with China's growth and demand for raw commodities, including crude oil. For observers of history, it was the prophecy of McBeth all over again. In fact, looking at the slope of the line in Figure 4.5, China's oil consumption clocked against its economic growth over the past 10 years is almost exactly the same as that of the United

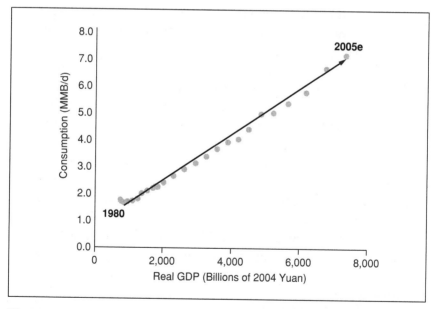

Figure 4.5   Accelerating Growth, Dependency, and Industrialization: China's Oil Consumption Cross Plotted Real GDP, 1980-2005e (*Source: Adapted from IMF World Economic Outlook Database and BP Statistical Review 2005*)

States back in the 1950s and 60s. Steep slopes (oil dependency factors) are hallmarks of rapidly industrializing economies.

Since 2002, the world economy has been firing on all cylinders. Asian economies recovered. Russia looked back on-track with reforms. Eastern Europe joined the E.U. and became progressively industrialized. Even Japan has shown positive signs that it is coming out of stagnation. In the United States, the bursting of the high-tech bubble and the aftermath of the World Trade Center attack is behind us, and the Enron fiasco has been dealt with and largely forgotten. The Bush tax cuts, low interest rates, rising home equity, and things like zero-rate financing for cars have served to free up disposable income in U.S. consumer bank accounts, allowing people to spend more freely than ever. In short, the world's economy has been growing in synch, every region at once. As a result, world GDP growth rose to 5.1 percent in 2004, which is one-and-a-half times the long-term average of 3.5 percent. The robustness carried through to 2005, which saw growth of 4.4 percent—all of this compounding against higher oil dependency factors. Predictably, the world's oil consumption grew aggressively. In 1997 the world consumed 73.7 million barrels of oil per day[4]. By 2002 the number had risen to 77.9 million barrels per day. But it's the recent three-year rise to near 86 million barrels per day, or a thousand barrels per second, that has catalyzed the pressure build.

Looking at Figure 4.6, you can see where all of the world's new oil demand is coming from. In 2005, Americans consumed close to 21 million barrels per day, three times as many as China, the next largest consumer. This means that small changes in the U.S. economy have a big impact on oil demand. However, the combination of supernormal GDP growth (10 percent in 2005) and a high oil dependency factor (over 90 compared to the world average of 37) make China and the rest of Asia a very large source—over 40 percent—of the 1.5 to 2.0 million barrels of new oil demanded every year.

Remarkably, the bulk of U.S. demand pressure comes from the biggest oil hog of all, the motor vehicle. The pie chart in Figure 4.7 details where all 21 million barrels per day of oil go in the United States. Transportation, especially road transportation,

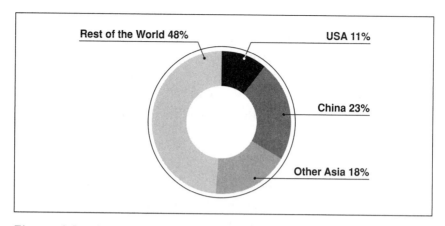

Figure 4.6   Sources of Incremental World Oil Demand: Percent by Region on a total 1.7 MMB/d 2005 over 2004 (*Source: Adapted from International Energy Agency Oil Market Report, July 2005*)

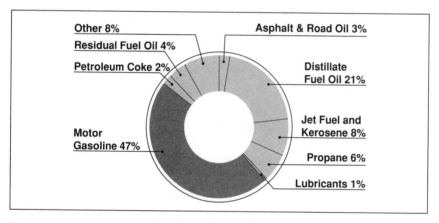

Figure 4.7   Carving Up a Barrel of Crude Oil: Percent Petroleum Products Derived (*Source: Adapted from U.S. Energy Information Agency*)

is clearly the biggest wedge in the pie. In fact, it's the biggest wedge of the pie in every nation, though some like the United States are more reliant than others. Understanding the influence of motor vehicles on the evolution of the energy cycle is paramount, because almost half the world's oil consumption ends up in a gas tank. How rapidly industrializing nations like China and India cope with increasing demand for mobility—in other words people buying cars—will have a big effect on how the world navigates and rebalances through the coming break point. Looking at the American experience provides sobering insight.

The automobile is an amazing device. We can barely imagine life without it, or how much our world has been changed by our reliance on it. Once a novelty reserved for the wealthy and acquisitive, after Henry Ford introduced the Model T on October 1, 1908, cars quickly became a necessity for the average consumer. Today, the network of supply chains that turns barrels of rock oil into useful motion is complex and mind-bogglingly inefficient. In Figure 4.8 you can see, in simplified form, how the original energy contained in a barrel of oil—100 units—converts down the supply chain from the "well to the wheels." As I mentioned in Chapter 2, by the time the rubber hits the road, only a dismal 17 percent of the original energy in a barrel of oil is actually converted to travel distance. Looking at the remaining energy available for use at the end of each conversion, you can see that the bulk of the 83 percent energy loss happens in the cylinders of the notorious internal combustion engine. Much of the energy loss is blown out the exhaust pipe as heat. Vehicles that stop and start a lot tax such an engine even more, taking the energy losses even higher, therefore pushing fuel economy lower. Contrary to popular belief, fuel economy is not all about the admittedly inefficient internal combustion engine. Weight compounds the issue of fuel economy too.

Oil-consuming vehicles in the United States break down into three broad categories: heavy trucks, light trucks, and automobiles. The division between the categories is mostly by weight, and somewhat by function. Anything greater than 10,000 pounds with more than two axles is a heavy truck. Pickup trucks, SUVs, and vans that range between 3,500 pounds and 10,000 pounds are in the light truck category. Anything leftover like cars, sports cars, and minivans are considered automobiles. For a sense of scale, the subcompact Toyota Echo treads lightly at 2,300 pounds (1,045 kg). A Chevy Silverado 4X4 or Hummer H2 carves up the pavement with 6,400 pounds (2,910 kg) of brawn. Automobiles and light trucks—the kind of vehicle you are likely to have in your own garage—are collectively referred to as "light vehicles." Currently, there are about 230

PROCESS	PROCESS DESCRIPTION	ENERGY BALANCE (units)
Crude Oil Refinery	The chemical energy in crude oil is converted into gasoline, which is a more refined source of chemical energy. The gasoline is then transported to gas stations. In this process about 83% of the original energy is preserved.	**100**
Gasoline Combustion Engine	An internal combustion engine converts the chemical energy resident in gasoline into rotating mechanical power. At best a gasoline engine is 35% efficient, so only 29 of the 83 units left over make it to the gearbox.	**83**
Mechanical Power Gears	Rotational energy from the crankshaft of the internal combustion engine is passed through a series of gears and mechanical processes until it finally turns the wheels. Frictional forces cannibalize about 35% of the 29 remaining units.	**29**
Rotating Wheels Road Traction	The wheels are turning, but the rubber has to hit the road. More frictional heat is lost in the process of gaining traction. In the end, only about 17 units of energy in a barrel of oil actually go toward the useful work of moving a vehicle.	**19**
Motion		**17**

(Energy Supply Chain)

Figure 4.8   From the Well to the Wheels: Accounting for the Road Transportation Energy Supply Chain. Note: Numbers are approximate and depend on many variables, including vehicle type, driving conditions, and habits.

million registered light vehicles in the United States, and through the year they guzzle 140 billion gallons of road fuel, mainly gasoline[5]. Largely undeterred by rising gasoline prices, more and more people are trading in their cars for heavier, less fuel efficient models like SUVs and pickups. Figure 4.9 shows the evolution of the sales split between cars and light trucks.

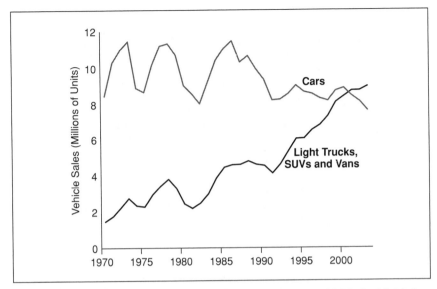

Figure 4.9   Annual U.S. Light Vehicle Sales, 1970-2003: By Vehicle Type (*Source: Adapted from U.S. Department of Transportation*)

Note that today's annual auto sales of around eight million units are no higher than in 1970, while the market appeal of light trucks has grown to exceed new car sales. Back in 1970, auto sales had 85 percent of the market's share; today that share has dropped to 45 percent. Utility, space, status, and perceived safety have gradually outsold fuel economy.

The shift toward heavier vehicles in the United States has had the effect of slowing down gains in overall fuel economy. Rapid improvements in fuel economy between 1970 and 1990 were mostly a function of improving engine technology, with fuel injection being one of many major innovations. Because weight is a dominant factor in fuel economy, the substitution of plastics for heavy metal parts also contributed to dramatic savings in gas mileage. By 1990, however, these gains were eroded by the overall shift to heavier vehicles and the nontrivial impact of increasing traffic congestion as people migrated out to the suburbs. Figure 4.10 shows fuel economy trends for each category of light vehicle, plus the overall average. Though gradual year-over-year improvements

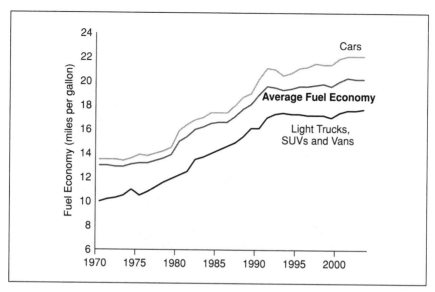

Figure 4.10   Average Realized Vehicle Fuel Economy, 1970-2003: By Vehicle Type and Overall Average (*Source: Adapted from U.S. Department of Transportation*)

have occurred for each category of light vehicle, overall fuel economy in the United States has pretty much hit a wall at just over 20 miles per gallon, even though CAFE standards of 27.5 miles per gallon have been fulfilled and there are car models available on the market that deliver up to three times the realized average.

Another big trend contributing to increasing fuel consumption has been the demographic migration from urban centers to the suburbs. As Figure 4.11 illustrates, a ramp up in commuting between the mid-1980s and late-1990s added 20 percent to the distance traveled by each vehicle—from 10,000 to 12,000 miles per year. Since the late 1990s the distance traveled by each vehicle may have started to level out, which notionally makes sense. After all, commuting any more than two hours each way to work is surely the limit for even the most dogged suburbanites.

Hand-in-hand with the migration to the suburbs has been increased traffic congestion. In stop-and-start traffic, there is no

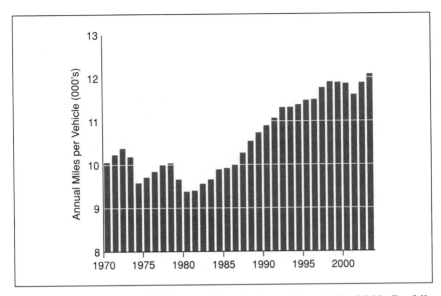

Figure 4.11   Average Distance Traveled per Year, 1970-2003: By All Registered Light Vehicles (*Source: Adapted from U.S. Department of Transportation*)

avoiding the tap dance between accelerator and brake pedals, which substantially decreases gas mileage over the commute.

Putting all the major trends and dynamics together, fuel consumption by American road warriors now is rising at a rough rate of one-and-a-half percent per year, or an extra 2.1 billion gallons. You can see this in Figure 4.12, where not surprisingly most of the growth is coming from the top wedge—the light truck segment.

But here's the kicker: it takes two barrels of crude oil to make one barrel of gasoline. Therefore, to carry on with the status quo—in other words, without any change to individuals' driving or buying habits—the oil industry needs to bring an extra 250,000 barrels of crude oil to market every year. And because U.S. oil production is declining, that oil will necessarily have to come from foreign sources. Now you get a sense of what's behind Figure 2.7, and why President Bush emphasizes the word "energy independence" in many of his speeches.

But U.S. energy independence is not all Americans should be concerned about. If we think like Reid Sayers McBeth and consider

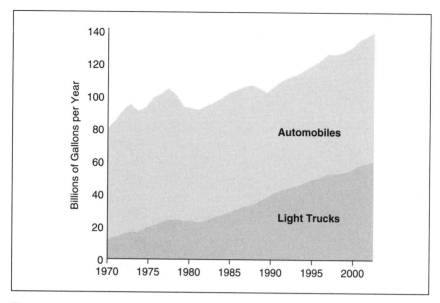

Figure 4.12   Growth in U.S. Road Fuel Consumption: by Vehicle Type (*Source: Adapted from U.S. Department of Transportation*)

that China's consumption pattern as a function of its economy repeats where the United States and other industrializing nations were at in the 1950s, there is almost no choice but to become alarmed. As 1.2 billion citizens of China seek the comfort zone of oil and its products, it's only a matter of time before their own break point is reached. New vehicle sales have been rising aggressively in China and are now running at between 350,000 and 400,000 per month, nearly four times the sales rate five years ago. Part of the rapid rise is attributable to China's admission into the World Trade Organization in 2001, at which time vehicle prices for domestic consumers fell due to the easing of import tariffs. Overall industrialization, retail gasoline subsidies, and wealth creation have been big catalysts too. The rising sales trend in Figure 4.13 is dramatic enough, but the bigger issue is where the limits lie. Only 8 out of every 1,000 people in China today own a vehicle. Contrast that to a global average of 120, and over 800 in the United States. With China's seemingly limitless potential to turn more

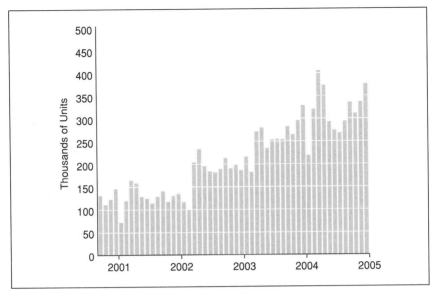

Figure 4.13   Monthly Vehicle Sales in China: All Cars and Light Trucks (*Source: Adapted from China Association of Automobiles*)

and more wheels, there's no doubt Reid Sayers McBeth, if alive today, would be feverishly penning another book and reiterating his original thesis that, "Petroleum today holds the front of the stage in a greater degree than ever before. As a wealth creator it never has been so fruitful as at present."

China's thirst and intensity for oil today is actually quite typical of a country in its early stages of industrialization and wealth creation. Thinking back to my Energy Evolution Cycle (Figure 1.1), China is resolutely in an early Growth and Dependency phase today, a phase that other nations have shown can last 20 or more years.

India gets airplay as the other big Asian tiger. Perceptually, there is a belief that India is as much of a problem as China when it comes to growing oil consumption. While its economy is growing aggressively, in fact India only consumes about 2.5 million barrels per day, growing by about 130,000 barrels per day, per year—or about one-third the volume of China. Why such a difference? For one thing, India's economy is still smaller and is much more of a service economy than China. Call centers and software companies don't consume nearly as much oil as steel factories and manufacturing plants. Another reason is that India is rich in natural gas reserves, which it has been actively exploiting, and is opening itself up to more liquefied natural gas imports.

Finally, there is another region of significant note that often goes overlooked. Let's call it the Rest of the World, or ROW. Comprising every region that is outside of the United States or Asia, ROW doesn't get a lot of attention, but it should. As our pie chart showed, 48 percent of all incremental oil demand comes from ROW—about the same as China and the United States combined.

A large sector of ROW, Western Europe, is like Japan. Its oil demand is not tied to changes in economic growth as in the United States, China, and India. On the other hand, Eastern Europe and Latin America have been quietly growing their consumption, and though they don't garner the fanfare that China does, it is imperative that we watch them. After all, the ROW is also a major contributor to world GDP and, as I mentioned before, the unique thing about the last couple of years is that the world's economic engine has been firing on all cylinders. We don't have to wait for China or the United States to slow down for oil demand growth to slow. Economic deceleration in any one of these regions may do the job.

If things keep going as they are in the world—an aggressively expanding global economy combined with greater oil dependency, both compounding on a growing base of oil consumption—you can get a sense that it all leads to unsustainable demand scenarios very quickly. And that, in short, is why we are keeping all of our storage tanks full, and why the market is contracting to buy oil at high prices five years out from now.

The Supply Challenge

So much for the demand story; now let's look at supply. For the last several decades, oil pundits have been prognosticating that the end of oil is near, but it seems as though each time someone has cried wolf, the giant oil companies of the world have been able to go further and deeper to find new reserves and prove the predictions wrong. Today some are crying wolf yet again, suggesting that soon we will not be able fulfill the one-to-two percent increase in oil consumption we need to keep up with global growth, or even maintain our current production levels. The debates are complex and the answers are not clear-cut. While I do not fall into the doom-and-gloom camp, I can tell you conclusively that feeding our growing thirst for oil is not going to be easy.

One of the larger oil fields found in the past 30 years was Hibernia, in the iceberg-cluttered waters of offshore Newfoundland, Canada. Chevron discovered that field in 1976. In the 1980s, I was part of a technical team that helped characterize the reservoir that eventually began producing in 1997. Initially it was thought that Hibernia held some 450 million barrels of oil, but today it looks like Hibernia's yield will be closer to a billion barrels. Better technology and a better understanding of the geology have helped to expand the reserve potential of Hibernia, but if the world was wholly reliant on Hibernia today, at our 1,000-barrels-a-second rate of consumption, we would drain this reservoir in a mere 11 days.

Unfortunately, there aren't many Hibernias turning up these days. Those that we do find are extremely expensive and highly risky to drill. In the industry, we refer to large multi-hundred-million barrel oil reserves as "elephants." As in the wilds of Africa, the world's elephant oil reservoirs are becoming extinct. Figure 4.14 shows a bar chart of annual oil discoveries since 1900. The height of each bar is billions of barrels found each year.

The peak was 1960 when elephant hunting was easy. From then on, major discoveries have become increasingly rare. As you can see, when prices spiked in 1979 and again in 2000, there was a brief resurgence in discoveries because oil companies had incentive to go to greater extremes to find oil. You can expect another blip up, but it's quite telling that the oil industry today only finds about 10 billion barrels per year as compared to 60 billion in the elephant hunting heyday. Given that the technology for discovering and extracting reserves has become dramatically more sophisticated, the

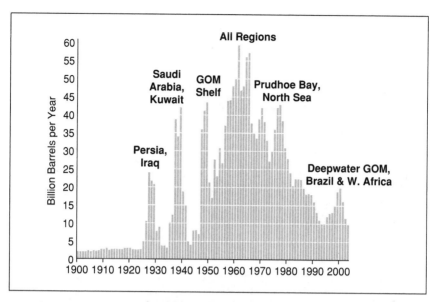

Figure 4.14   Total Volume of New Oil Discoveries Worldwide: By Year, 1900-2004 *[Source: Adapted from* Harper *(2003) and* Oil & Gas Journal *(2004)]*

lack of big discoveries today is testament to the fact that the world's remaining reservoirs are not as plentiful or as big as they used to be. As companies like Shell and BP go ever further offshore, drilling in deep waters off of the Gulf of Mexico, the North Sea, the coast of West Africa, or off Sakhalin Island in the northern Pacific, one can't help but reflect on the whale men of a century and a half before who worked those same waters in search of the sperm whale. Indeed, finding an elephant today is about as rare as sighting a sperm whale off Boston Harbor. Once again we are searching the ends of the earth for our insatiable energy needs.

Historically, oil was found by following seepages along cracks in outcropping rock formations. Exploitation of oil seepages probably goes back to prehistoric days. Early tribes in Europe, the Caspian, and North America all used oil from oil seepages to make heat, heal wounds, waterproof canoes, and soften leather.

There have been many colorful figures in the long history of oil discovery. George John "Kootenai" Brown, an Irishman who

Figure 4.15   Discovery Well at Oil City: By Cameron Creek, Alberta (circa 1902) (*Source: Glenbow Library and Archives*)

transplanted himself into the wild west of Canada, was an adventurer, packer, and mountain guide near Waterton in the picturesque foothills of the Rocky Mountains. With news of oil discoveries in Pennsylvania and Ontario spreading west, fortune hunters like Brown sprang into action, searching for supply. Though Brown was the first nonnative to find seepages in the wilds of western Canada, it was left to others with greater business sense to commercialize the product. By the early 1900s, entrepreneurs had brought capital and established drilling techniques to the West. After that the race was on.

Whether it was in Alberta, Pennsylvania, or Texas, as soon as one oilman struck oil, others would hasten into the area, secure land positions, and start drilling, usually into a different section of the same reservoir. At some point, the unlucky ones would come up dry, because their wells were out of strike range.

As the science of oil drilling became more evolved, oil men came to understand that an oil reservoir is not like a big cave underground. It's typically a rock layer that is porous. In the ideal scenario, it can be thought of as a hard, oil-soaked sponge or solid Swiss cheese with all the holes interconnected and filled with oil. When a rig drills into the rock layer, the oil is free to flow through the interconnected pores and up the bore hole. Depending on the type of reservoir and its depth beneath the surface, there can be enough pressure to push the oil up the borehole to the surface, producing a rush of oil. In extreme cases, the oil comes rushing out of the hole like a vertical fire hose. This is the stereotypical "gusher" of oil lore, made famous at Spindletop, Texas, in 1901. It's much more common that there isn't enough pressure to push the oil up to the surface, or that the pores are not well connected, or both. Most of the time the oil needs to be pumped up from the bottom of the hole by a classic pump jack bobbing up and down, or some other pumping mechanism.

Oil is found all around the world, in all sorts of geological settings—deep, shallow, porous, sandy, gravelly, salty, offshore, onshore, and so on. In terms of calculating the oil that is available to us, it's important to distinguish between reserves and production. Reserves refer to how much oil is in a reservoir that engineers think can be recovered. Production rate is how fast the reserves are pumped to the surface and pushed into a pipeline or tanker. Once a company has established an oil strike, estimating the amount of reserves available is a tricky endeavor. The obvious problem is that you can't see underground. It's also not usually clear how much of the oil in those interconnected pores can actually be recovered. There are considerable physical limitations to how much can be extracted, but a lot depends on price too; the higher the price the more an oil company is willing to spend on recovery. So, economic reserves that publicly traded oil companies report in their financial statements differ from actual physical barrels in the ground that are called 'oil in place.'

The statistics for recovery are actually not very good. Where the pores are tight and the oil is of low, viscous quality, only a dismal 15 percent of the reserves may be recovered. In regions like Saudi Arabia that have premium light sweet crude in porous reservoirs, the recovery can be up to 50 percent. Although that sounds good compared to 15 percent, it's fairly shocking to realize that at best only half a find is recoverable. On average, the global recovery factor is probably not much better than 30 to 35 percent. In other words, two-thirds of all oil discoveries are left behind in the ground. And if the reserves are developed recklessly, the reservoirs become damaged and even less oil is able to be pumped out. There are many high-tech tools today that can help find new reserves and enhance recovery, but there is still considerable uncertainty as to how much oil can actually be brought out of the ground, especially with newly discovered reserves.

Accordingly, there are two angles to the, "Are we running out of oil" debate. The intuitive angle is that we are indeed running out of reserves, but that's just simply not the case. There are billions of barrels of oil reserves remaining on the planet. The real issue is that we are running out of reserves that provide enough economic incentive to produce with today's technology and in our current geopolitical atmosphere.

It's not a matter of running out; it's really more about the rate at which the oil industry can pump the oil out of the reserves. After all, the oil is of no use to us if it just stays trapped in the rocks, because we can't use it in our cars, to heat our homes, and so on. If the oil suppliers can't keep up with the growing demand, it may amount to the same thing as running out of oil, at least in the short term.

At the moment, the world's oil industry is pumping out crude oil at about the same rate we're consuming it—1,000 barrels a second. A recycling of profits through massive capital investment in drilling, pipelines, facilities tankers and refineries is required to keep increasing the rate to meet tomorrow's needs. Because it's becoming harder to find the shallow reserves, it's getting more and more expensive to increase the rate of production. Across the globe, oil reserves in many regions are maturing. Telltale signs of a region's maturity are increasing costs to bring out the new barrels of oil as well as a peak in productive capacity. In other words, we know a region is maturing when the rate of production cannot be increased no matter how many more wells the industry drills. Such is the case with the United States, the nation with the longest history of crude oil production. Production out of the once-substantial U.S. crude oil reserves peaked in 1970 at 9.6 million barrels per day. In 2004, 145 years after it first began, U.S. production was down to 5.4 million barrels per day.

Those who subscribe to a theory called "Hubbert's Peak" suggest that the entire world's oil reserves are maturing and that production has, or is very close, to peaking. Followers of Hubbert's doctrine—and there are indeed some very fervent believers in his theory—say that in the next half dozen years or less, the world's oil industry will be unable to increase the production rate out of our existing and new reservoirs. In other words, Hubbert's disciples think that world oil production has peaked at about 1,000 barrels per second.

M. King Hubbert was an American geophysicist working at Shell, who predicted in 1956 that U.S. oil production would peak in the early 1970s. He was bang on in his Nostradamus-like prophecy, which is why his work attracts so much attention today. Hubbert's supposition follows the well-documented dynamics of how humans tend to use up natural resources. Whether coal, copper, or oil, natural resources are exploited in a rate pattern that is remarkably close to a "bell-shaped" curve. First slowly; then the rate of production increases rapidly; then it peaks; then there is a rapid, symmetrical decline into maturity, followed by a slow demise. The area under the bell-shaped pattern is the total volume of the resource that can be recovered.

Depending on who or what you believe, over the past 300 million years, the natural forces of geology acted to create somewhere around 2.2 trillion barrels of conventional oil on our planet. Figure 4.16 shows how the world's rate of crude oil production has been increasing over time (solid line) and a "best fit" bell-shaped pattern representing the 2.2 trillion barrels. Estimating these total recoverable reserves, and fitting the bell curve to the historic production

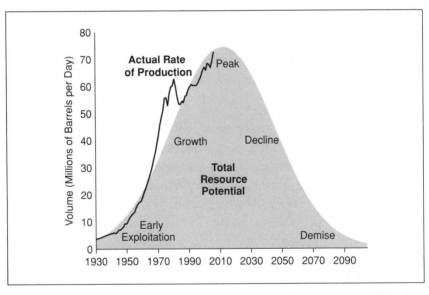

Figure 4.16   Hubbert Type Peak Oil Production Analysis: Global Production History and Best Fit Bell Curve (*Source: Production adapted from U.S. Energy Information Agency, bell curve from ARC Financial*)

data, Hubbert followers point out that we are either at or very near the peak rate of global oil production.

And guess what? They're right—but with some important caveats. First there is enough subjectivity in the analysis that the peak could be next year, or 20 years from now. Much of the error in placing the peak originates from determining what the true volume of total oil reserves are under the bell curve. Of course we will never know the true volume until the last drop is pumped out. Detractors of Hubbert's peak correctly point out that the amount of economically recoverable oil left on the planet depends on price. The higher the price of oil can go, the more the oil industry will be able to scavenge for oil reserves at the far fringes of our planet.

But not all oil is created equal. There is heavy oil, light oil, bitumen, oil sands, oil shales, and other sources. The world's refineries are not fitted to process all of the various kinds of oil available; each refinery has its own preferential slate. As price rises the industry starts chasing after heavier and heavier grades of "nonconventional" or "secondary" sources of oil, like the Canadian oil sands in Alberta, and the U.S. oil shales in Utah and Colorado. We will require more and more specialized refineries that can chemically convert those oils into lighter grades that we can use.

The notion that we will need to seek out heavier grades of nonconventional oil when we reach peak production is not new. Predating Hubbert's work by 30 years, Sir John Cadman, then chairman of Anglo-Persian, hypothesized on November 2, 1927 that: "Very many years must elapse before natural petroleum resources will be unable to meet the greater part of the world's requirements. Of course, the time will eventually come when the world may have to look for a great part of its supplies from secondary and synthetic sources, but he would indeed be an optimist who imagined that—on the reaching of such a stage—prices would remain as low as those existing in the past."[6]

Almost 80 years have passed since Cadman's address. It's enough time to have seen many of the world's prime reservoirs mature. For light sweet crude, the evidence strongly suggests that we are very close to Hubbert's peak, and that we have reached the stage

Cadman speaks of where we must start exploiting secondary and synthetic sources to prolong the onset of the overall oil peak and expect oil prices to go up.

Reservoir maturity relates to the physical phenomenon of natural production declines, one of the biggest challenges facing the oil industry today. Left to its own, the rate of production from any given oil well will decline over time. For example, a well that starts out producing 100 barrels per day in its first year, may only produce 75 barrels per day after a year, and 56 barrels per day the year after. This well is said to be declining at 25 percent per year. Decline rates vary throughout the world depending on geologic circumstances and how the wells have been engineered, but you can't get over the physical reality that every oil well loses productive capacity over time. To start, a typical oil well's decline rate is high, but gradually it stabilizes to less than 10 percent per year as it matures.

Estimates vary, but the overall average global decline rate is now somewhere between five and eight percent. The implications are profound. If oil companies do not spend money on new drilling, production in 2006 will decline by 4.3 million barrels per day in 12 months, assuming a conservative five percent decline. In two years, 2008, we would be down to 77.6 million barrels per day. So after a mere two years, the world's production would regress back to where we were 10 years ago. If you're more pessimistic and choose an average eight percent decline those production numbers would be even grimmer.

To put the world's oil decline rate in perspective, 4.3 million barrels per day is about two-and-a-half times Iraq's production right now. In other words, right now the world's oil industry has to find the equivalent of two-and-a-half Iraqs every year just to maintain today's production levels!

Of course, it's not good enough just to offset the five percent decline. Remember, we have to *grow* production every year. As long as the world economy grows, demand for oil grows. Right now, because of the high oil dependency exhibited by China and the big engine called the United States, production has to increase by about two percent per year. We are thus faced with a situation in which we are taking five steps backward for every seven forward.

Unfortunately, even if we could offset our growing demand needs and our declining production, that still wouldn't be enough. We need a buffer of oil, beyond our consumption needs, in order to get by. Think of the world's oil supply chains as one big manufacturing system. What manufacturing system do you know that operates full out at 100 percent, 24/7? Every assembly line requires downtime for maintenance. It requires "spare capacity" to be able to handle unforeseen events.

In today's oil industry, the only spare capacity exists in OPEC countries where oil production actually exceeds current demand or "call." Non-OPEC exporters like Russia, Canada, Norway, and Mexico are all assumed to be fully called and therefore have no excess production. So it is only OPEC's slim margin of excess production that stands between us and a global oil crisis, and that's what makes them the most powerful group of oil producers in the world.

No energy supply chain can operate at 100 percent capacity. Coal-fired power plants typically run at a utilization rate of around 75 percent. Nuclear power plants go harder at 92 percent. Wind turbines are lucky to operate at 35 percent for the simple reason that the wind doesn't always blow. Running any manufacturing or energy system at 100 percent capacity is unrealistic and leaves no room for error; yet, today, we are practically doing that when it comes to crude oil supply.

Figure 4.17 characterizes the major issues. From 1970 through 2005, the world's oil companies have produced 868.3 billion barrels of oil, represented by the area up to the vertical white dashed line. Now, let's spring forward to 2020. Assume for a moment that the world's oil companies were to stop spending money on finding new oil and just "blew down" their reservoirs, emptied their cupboards if you will. Another 311.5 billion barrels would continue to come out of the ground, represented by the dark grey area between 2005 and 2020. But note how the rate of production goes down exponentially, courtesy of our five percent yearly rate of decline.

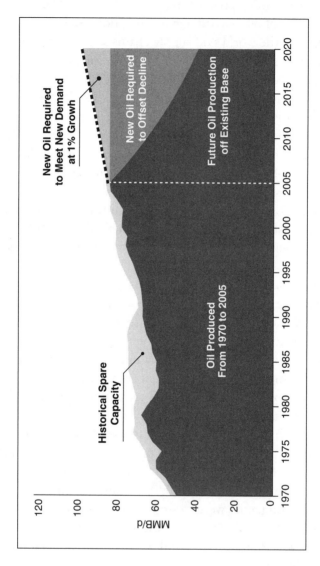

Figure 4.17 The Challenge of Supplying Global Oil Demand Growth: Historical and Future Components of Supply, Demand, and Spare Capacity (*Source: Adapted from U.S. Energy Information Agency, the International Energy Agency, and ARC Financial*)

To offset the five percent decline and still tread water at a 2005 rate of production, the oil industry must find and extract another 146.8 billion barrels, represented by the light grey area.

But world oil demand is growing every year, not staying level. The grey wedge represents the amount of oil needed to be found and produced to fulfill a one percent growth rate out to 2020, equaling 41.4 billion barrels. This number is actually quite conservative when you reflect on how much growth we could potentially see in China, other parts of Asia, Eastern Europe, and the United States. The dashed line extending from 2005 to 2020 is really our minimum demand challenge.

Finally, the layer covering everything on the top of the grey area, from 1970 to 2005, is the historical level of spare capacity—the world's safety blanket. Today, the blanket is exceedingly thin at around two million barrels per day. This means that the world's oil supply chains are operating at over 97.5 percent capacity, leaving us with very little error margin for the known unknowns—downtime due to natural disasters, accidents, unforeseen maintenance, terrorist attacks, geopolitical tensions, and so on.

No wonder the markets are sensitive, a condition always reflected in price. As supply and demand tighten, the spare capacity blanket thins out, people start hoarding, and price rises exponentially. Often there is a trigger point, a spare capacity level below which buyers and sellers in the marketplace start to get panicky. Price becomes very volatile and sensitive to even the smallest piece of negative news—a labor strike, a hurricane, a bombing, a refinery fire, and so on. We crossed under that trigger point, about 3.5 million barrels per day, back in 2003, and have been well under it ever since. A more substantial spare capacity buffer is needed if we want to keep the economy and the financial markets assured that we won't experience an oil shortfall. Until that time, the current hoarding mentality will dominate market actions, and prices for crude oil and petroleum products will remain volatile and high. Conditions for a break point are ripe.

In the 1990s, when oil prices had risen or supply had been, in some fashion, made temporarily insecure, the U.S. government pulled out the pacifier of the Strategic Petroleum Reserves, or SPR. The SPR was originally created by the United States for strategic military purposes. In case the world's oil supply was threatened, the United States would have a backup reservoir available to serve its military needs. Interestingly, the concept of the SPR was actually started by Woodrow Wilson during the time of the post-WWI scramble for oil assets when he set aside a significant supply of crude oil for the U.S. Navy at Teapot Dome, Wyoming. But it was the break point of the 1970s that really awakened the need to hoard a big reserve. From a few million barrels in 1977, the American SPR grew and leveled out to near 600 million barrels by 1990.

In 2001, the Strategic Petroleum Reserve made the news again for a different reason. Another build phase was mandated soon after the terrorist attack on the World Trade Center. In 2005, the SPR was filled up to meet President Bush's target of 700 million barrels. In the past, President Bush has said that he wants to see

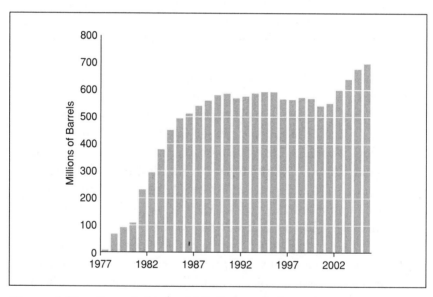

Figure 4.18   Growth in the U.S. Strategic Petroleum Reserve: Oil and Inventory at Year-End (*Source: Adapted from U.S. Energy Information Agency*)

the SPR built up to one billion barrels. That sounds like a lot, but it's only enough capacity to substitute foreign imports for three months if it came down to a megacrisis. To build up to a billion barrels of strategic reserve will require additional storage infrastructure; more importantly the hoarding will take valuable oil away from commercial use.

And why should we believe that the United States is alone among the world's nations in urgently filling up its SPR? Though statistics are difficult to uncover, it's almost assured that China and India are starting to stockpile their own strategic reserves. Wouldn't you, if you looked at oil supply today and saw that your nation will need to:

1. Offset the decline rate just to keep up the current production level;
2. Increase production to meet growing demand; and
3. Maintain a spare capacity buffer to keep the commercial markets confident that you won't run out of oil suddenly because of unforeseen calamity?

Hoarding is not just a reaction that auto drivers have when the availability of gasoline becomes uncertain; nations are also quick to hoard when they sense a coming break point.

Geopolitical Pressures

As demand for light sweet crude continues to grow at an aggressive clip, and the world's oil industry is finding it increasingly expensive to supply, the pressure in our energy cycle is building. Volatile and rising price is a blaring signal telling us that something is going on. It doesn't get much simpler than that, but it does get more complicated. There are many other forces than supply and demand contributing to the pressure in our oil supply chains today, exacerbating an already difficult situation and accelerating our speed to the break point. Referring again back to our Energy Evolutionary

Cycle in Figure 1.1, these are the forces acting to build pressure from outside the normal evolutionary cycle—things like geopolitical tension, environmental issues, social forces, and the dynamics of business and government policy.

In terms of geopolitical tension, the oil industry is now at the nexus of uncertainty. While we often think of global tension these days in terms of war and terrorism, there are market and policy forces beneath the surface adding more pressure than most of us are aware. In the 6 to 12 months prior to the U.S.-led invasion of Iraq in 2003, the price of oil was thought to be between $5 and $10 higher than it should have been due to tensions in the Middle East. This conjectured premium on oil price was dubbed the "war premium" because many analysts felt it was the incremental price the market was paying for the uncertainty of how the diplomatic sparring between the United States and Iraq would play out.

Iraq was invaded, of course, and the rapid advance toward Baghdad shaved close to 10 percent off the price of oil almost overnight. In the days following "Mission Accomplished," hopes for peace and rapid reconstruction were high. On April 9, 2003, when Baghdad fell, Vice President Cheney predicted that Iraq would be producing, "on the order of two-and-a-half, three million barrels a day," by year end[7]. That would have been more than a full recovery of Iraq's pre-invasion production. The war premium not only disappeared; it turned into a "war discount" with Saddam Hussein deposed and important oil supply chains originating from the Gulf secured.

Today, any discount is history, and there is no doubt that the war on terrorism, the ongoing conflict in Iraq, and the specter of more conflict in the ever-tense region of the Middle East have the oil markets on edge. More than usual, the markets are concerned with the issue of concentration of supply in such a vulnerable region of the world. In fact, there's good reason for that concern.

The Middle East is the world's dominant supplier of oil. Just less than 20 percent of the world's oil supply passes through the mouth of the Persian Gulf via the Strait of Hormuz. While not all of the world's supply travels that narrow waterway, almost all of the world's spare capacity is behind that opening, including the great mother of all oil-supplying nations, Saudi Arabia. Countless books, articles, and academic papers have been written about Saudi Arabian politics. Opinions on what will happen in the Kingdom of Saud over the next 10 years range from status quo to apocalypse. Those very concerns are a sign that there is too much of one of the world's most vital commodities in one place, that we've put all our oil in one basket, or barrel.

People who manage investments for a living refer to such a circumstance as a "concentration of risk." The answer to such a concentration of risk is to diversify your investment portfolio. But in the case of oil, where supply chains are currently operating at close to 98 percent of capacity, how are we going to diversify our consumption? There are 192 countries in the world[8], and nearly all are dependent on oil. On the flip side, only 30 countries produce oil of any significant quantity, and only 17 of them are exporters of oil greater than 500,000 barrels a day. Geography and politics limit choice. For most consuming nations, diversification to different suppliers is limited to about a dozen countries, few of which would make it onto any preferred supplier list. A high concentration of the world's oil supply is not only buried under rock, ocean, or sand, it is buried under layers of corruption, political risk, and capricious authoritarianism.

The market senses this concentration of risk and directly translates it into price volatility. Some oil-producing countries like Nigeria and Indonesia live under a perennial cloud of out-of-control crime, insurgency, civil war, and broader armed conflict. Drug lords, guerrillas, and militants hide out in the jungles of places like Colombia, threatening oil workers. Zealots in Iraq routinely blow up pipelines. Expatriate workers that agree to work in these countries today command huge salaries and bonuses for risking their lives on a daily basis. It all adds to the cost of doing business. Every time Hugo Chavez, the populist president of Venezuela, delivers a scathing speech against George W. Bush and the United States; every time there is a terrorist attack in Saudi Arabia; every time there is a strike, ethnic conflict, or incident of pipeline sabotage in Nigeria; every time there is tension with Iran over terrorism, atomic energy, or Islamic fundamentalism, the markets are reminded of the concentration of risk and react accordingly. If spare capacity is really tight, as it usually is in the winter months, the price volatility is amplified. In the market's view, if there is anything worse than high prices, it is price volatility. Industrial consumers, trucking companies, and airlines do not like uncertainty in price because they cannot plan their budgets. Entire sectors of the economy are thrown into doubt as a result, and a domino effect can easily follow.

One hope for an alternative to diversify—like it was 100 years ago—is Russia. But perceived political risk factors in Russia are also significant today. No doubt Russia will continue to supply an increasing fraction of the world's oil demand, but westerners who have been pinning hopes on Russia as a safe and secure source of future oil need to be reminded of the regional history. Russia's vast oil reserves have been under state control a lot longer than they have been open for capital investment. The nation has a history of nullifying independent oil company interests through nationalization and using oil as a strategic tool to exert influence on the world. More broadly, we should not forget that actions by countries and corporations to control the world's greatest oil reserves are nothing new—these actions have been going on since before WWI.

Concentration of risk, which includes corruption and all the components of political risk, manifests itself financially in terms of higher return on investment requirements, also known as hurdle rates. The higher the risk, the higher the return necessary to make the investment worthwhile, the greater the hurdle. Large oil companies that invest in Canada (a substantial exporter with little political risk) often use a hurdle rate of around eight percent after tax. This means that for every $100 they invest, they expect a minimum net return of $8. No independent oil company will risk its people, equipment, and money to go into places like Venezuela, Libya, Indonesia, or Russia for a measly eight percent return. Anecdotally, an unwritten requirement of at least 20 percent return (and rising) is required to offset the kind of uncertainty that comes with doing business in these countries. The combination of increasing concentration of risk and political risk means that the hurdle rate keeps going up around the world. The higher the hurdle rate, the higher the long-term price of oil has to be before independent oil companies have incentive to go find and develop reserves. Put another way, the increasing risk dynamics no longer support viable, free-market economics at the historic $20 a barrel. What is supportable? With so many fluid dimensions in this pressure build period, it's hard to say what the threshold price of oil has to be now before oil companies and their workers have incentive to overcome all the risks and bring new barrels of oil to consumers. My readings, calculations, and anecdotal discussions suggest that there is little incentive now below $40 per barrel, and even that may be too low.

But the world's oil industry is not all about independent, free-market, non-state-owned oil companies like ExxonMobil, Chevron, Shell, and BP. In fact, the influence of the independents has waned over the decades and only one, Lukoil of Russia, makes the top-10 list of oil companies as measured by reserves. The top 9, led by Saudi Aramco, are all 100 percent state owned. For a sense of scale, ExxonMobil is twelfth on the list with one-twentieth the reserves of Saudi Aramco. State-owned oil companies representing both producing and consuming nations are the norm around the world, and all are getting more aggressive. It's like the post-WWI great scramble all over again.

The Great Scramble Returns

China became a net importer of oil back in 1993, but the steep ramp-up in oil imports really began in 2002 or 2003. Currently, China requires almost 4.0 MMB/d from sources abroad. The pattern of increasing dependence is reminiscent of the United States back in the early 1970s—and a major contributor to the pressure that is building towards the imminent break point. The situation is not limited to China. In the face of rising demand, especially in Asian countries, oil companies around the world are once again

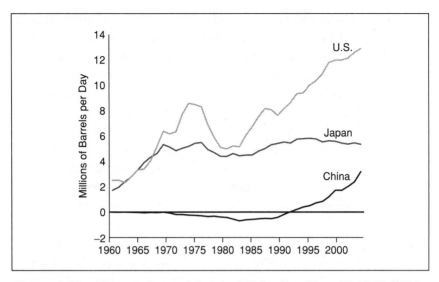

Figure 4.19   Comparison of Crude Oil Import Trends, 1965-2004: United States, Japan, and China *(Source: Adapted from BP Statistical Review 2005 and ARC Financial)*

chasing after the world's scarce reserves. Some of the old names, like ExxonMobil, Shell, Chevron, and BP, are still key players in this renewed Great Scramble. But significant other companies have emerged like China National Petroleum Company, Oil and Natural Gas Corporation, Sinopec, Petrobras, Lukoil, Petronas, and many others.

With more players and fewer opportunities, the price for the right to explore, develop, and bring oil to the market is going through the roof in places like Libya, Kazakhstan, and even Canada. It's akin to a real estate boom with underground properties. Complicating this Great Scramble is the age-old rivalry between independent oil companies (IOCs) and national oil companies (NOCs). Eighty years ago, independents like Standard, Shell, and Gulf pushed their weight around and dictated terms to technologically and financially devoid regions of the world. The sophistication of the IOCs was essential to coax oil out of the ground and to the market for the benefit of all parties. Today, the NOCs are a lot bigger, savvier, and more aggressive than they were in the

1920s. To top it off, they have distinctive competitive advantages over the IOCs.

Most large IOCs are publicly traded and must fully disclose their finances and activities to a large base of shareholders that are expecting double-digit returns on their money. Sarbanes-Oxley, legislated in the post-Enron regulatory climate, has amplified the need for disclosure and heightened shareholder scrutiny. In addition to, and perhaps in outright conflict with, this added scrutiny, IOC shareholders also expect their companies to be outstanding citizens of the world, mindful of environmental and human rights issues in whatever country they are operating. While such behavior is to be applauded, many rival NOCs don't feel the need to share the same exemplary value systems.

For example, Canadian-based independent Talisman Energy was morally browbeaten out of Sudan by shareholders and lobby groups concerned about human rights violations. Taking Talisman's place, one of India's national oil companies, OGNC, moved into Sudan quickly, feeling little such altruistic pressures.

Another factor creating the uneven competitive playing field for oil concessions is that many national oil companies don't have quarterly shareholder pressures that place near-term results ahead of long-term vision. For that reason, NOCs, are more apt to outbid large IOCs on key oil concessions. And why not? NOCs owned by high-growth nations like China, Malaysia, and India have a vision to secure access to long-term oil supplies. For NOCs, serving their energy-hungry mother countries is their crucial shareholder objective. In other words, security of supply trumps near-term profitability for state-owned oil companies representing large consuming nations.

Because oil production, and especially oil export capacity, is so concentrated in the hands of a few nations, supply-side government policies in those nations can have a big impact on the pressure buildup. The most obvious policy structure at play in that regard is OPEC, a collection of 10 nations that have historically had huge influence in controlling the world's marginal oil supply and hence its price. But in a world where the old rules of oil pricing no longer seem to apply, we must question how much influence OPEC has now.

Much of OPEC's influence has been predicated on being able to manage spare capacity. Collectively turning taps on and off as a cartel has been a mechanism that has generally worked to control prices in the past, or at least when the often unruly group of nations has had the discipline to act like a cartel. In the past OPEC introduced a price band mechanism to guide prices[9] between $U.S. 22.00 per barrel and $U.S. 28.00 per barrel by manipulating the balance between supply and demand. It all worked well in principle until 2004, when the whole concept of production quotas started losing relevance. As the world started to call on every barrel of oil that OPEC could produce, prices shot way beyond the upper band limit. Today, the taps are still nearly wide-open in all the OPEC nations, so there is not a lot of ability to control price on the down side. In particular, there is little spare capacity to buffer the world's growing demand during the high-demand season between October and March.

OPEC production quotas are now only worth noting because they may suggest a floor price during slower periods of demand, which at some point is inevitable. Otherwise, OPEC's stipulated targets have now become largely irrelevant. An OPEC-member country like Indonesia, for instance, has no spare oil left to export. Even Venezuela has had its quota raised at a time when it can't produce enough to meet its prior quota commitments. OPEC, especially member countries like Saudi Arabia, can regain control of the taps only if they continue to invest substantial capital over the next several years to turn reserves into production. Holding hands with President Bush in April 2005, Prince Abdullah committed an investment of $50 billion to keep developing reserves. Other OPEC countries are following suit by committing capital through their national oil companies or opening their concessions to foreign independents. The terms are tough, but today's high prices are catalyzing the action.

It's difficult to get an objective read on how fast reserves can be turned into production with the proposed levels of investment. Naysayers believe that older elephant oilfields within places like

Saudi Arabia have matured, and do not think that OPEC can add much more productive capacity regardless of investment. On the other hand OPEC, especially Saudi Arabia, is confident of growing supply, albeit on their own terms. There is yet to be a consensus as to where the truth lies, though it's probably somewhere in the middle (as is often the case). But the important point is that we appear to be in a new and rare era in the history of oil where there isn't a clear institution that seems to be in control of oil supply, and therefore price. Maybe that is OPEC's intent. If that's the case, then OPEC's measured, and seemingly relaxed investment pace today, is in fact a new form of price control in itself.

The importing nations of the world, scrambling again for reserves, don't have much choice in choosing suppliers that are low on the political risk spectrum, but one such choice is Canada. Contrary to popular belief, Canada, not Saudi Arabia, is the largest supplier of oil to the United States. What's more, Canada has a lot of productive capacity for the future. The Western Canadian Sedimentary Basin (WCSB) still ranks high as a prolific source of conventional crude oil and natural gas. Canada's nonconventional oil sands are now recognized as holding the second largest source of recoverable oil in the world, after Saudi Arabia. And unlike Saudi Arabia, there is no debate that capacity can be raised by steadily increasing oilsands output. Frontier regions like offshore Newfoundland are more icing on the cake. Although Canada's up-front costs of finding, developing, and producing new barrels ranks at the high end of the spectrum, by any metric it has become an energy superpower. Canada is a country that does not live under a perennial cloud of civil war or armed conflict. Its highly developed capital markets and fiscal regimes catalyze entrepreneurial exploration. What's more, Canada is a marketer's dream, positioned right next to the largest consumer of oil in the world, the United States.

Geopolitical events around the world are making Canada one of the *de facto* low-cost producers of new oil barrels. Is it any surprise that China made its third deal with the Canadian oil industry in 2005? The Sinopec Group entered into a $85 million dollar deal in return for a 40 percent interest in Synenco's Northern

Lights oil sands project. According to Synenco, the project is anticipated to start producing 100,000 barrels per day of synthetic crude oil by 2009 or 2010. Speculation is ripe that Sinopec is planning to see some of that production make it to the West Coast and on to China. Although the China National Offshore Oil Corporation's $19 billion takeover bid for U.S. oil company Unocal in 2005 was the first shot across the bow for U.S. politicians suddenly concerned with security of supply, the warning signs had been noticeable for some time in Canada, right in the United States' own backyard.

The Great Scramble for oil is back, and places like the Canadian oil sands are at the fore of attracting much attention and capital. Sir John Cadman's prophecy has arrived; the time has come when the world has to look for a great part of its new supplies from secondary and synthetic sources, places like the vast Canadian oil sands. And as he prophesized, it's only the true optimists that believe that prices are going to remain as low as in the past.

Numerical Risk

Another business factor contributing to keeping oil price volatile and adding to the pressure on the energy supply chain is the dearth of critical industry data.

Currently, there are several billion dollars worth of oil sales transacted every day. Compare that to Wal-Mart, one of the largest corporations in the world, which has sales averaging about $820 million dollars per day. Wal-Mart issues press releases about its business dealings almost daily and provides in-depth, audited, and certified disclosure of its financial activities. Conference calls and guidance provide analysts with a pretty clear picture of what's to come. Investors that buy and sell Wal-Mart shares make their trading decisions based on well-documented, independently audited information. The market for Wal-Mart stock is generally transparent, meaning that investors have good visibility as to what's going on.

Compare that with the oil market. Current information on oil supply, demand, and storage coming out of the marketplace is often infrequent, lagging, and incorrect. There is one official repository of information for the global oil industry and that's the International Energy Agency (IEA), based out of Paris. Founded during the oil crisis of 1973-74, the IEA today acts as an advisor for 26 member countries. It offers many research services including gathering data on supply, demand, and inventory levels from countries around the world, tabulating the information in their *Monthly Oil Report*, which is typically published in the second week of each month. The report is influential and often causes ripples through the oil markets. Arguably, it's too influential because it's the only publication that really shapes the market's opinion. It's as if shares of a company with four times Wal-Mart's sales were traded on the opinion of only one analyst!

The data published by the IEA is always susceptible to revisions because the supply and demand volumes don't balance easily—in other words, supply minus demand rarely equals what is withdrawn or put into storage. There is always a balancing item, or fudge factor to account for all the oil produced, consumed, and stored. The corrections can sometimes be very large, and can accumulate over time. In the late 1990s, when oil prices were really low, the accumulations were so large it led analysts to question where these "missing barrels" were going. Was the market really oversupplied, or was the data just plain wrong? Poor data integrity is not necessarily the IEA's fault, because most of their information comes from the bureaucracies of the countries they are polling. For one thing, it's not realistic to think that impoverished countries can provide the same level of data quality as say the United States, Canada, or the United Kingdom, which spend millions of dollars gathering industry stats. Also, many nations are protective of their numbers, with their state-owned oil companies shielding the detail behind their supply and demand numbers from public scrutiny. It's hard to get to an absolute truth, because there is no Sarbanes-Oxley-type act governing the numbers that the IEA gathers and tabulates.

Despite the thick data fog, it's common for analysts to use IEA data as a base of information, adjust the numbers to reflect their knowledge and intuition, and then try to extrapolate supply and demand based on information that can be gathered from other sources. Take the situation in China as an example. It's imperative to get a sense of China's GDP and oil consumption, because the region is so important to world demand. Yet the official numbers that China publishes are often lagged by several months, and many outsiders question the accuracy. For other countries, GDP numbers, oil consumption stats, and other pertinent data to help estimate the forward fundamentals are reported only annually and are usually lagging by a year. Data for 2004 is often not released until the end of 2005, not much use in a rapidly changing oil market.

There are several private research and consulting firms that make pretty good estimates of what's going on in the world of supply and demand. Unfortunately, their root source often traces back to the IEA too. More importantly, these firms are private and sell their data and analysis to selective clients, so the data is much less visible to the broader oil markets. Good sources of free public information include the United States Energy Information Agency (EIA) and the BP Statistical Review. The data from these two agencies is as good as any, but timeliness and frequency of information remains a big issue. The BP data, though comprehensive, is released once a year. The EIA offers weekly stats on the U.S. industry, but global updates are far less frequent.

The upshot is that the oil markets are often trading in dim light, and such uncertainty translates into greater volatility. Because oil is such a vital commodity, the tendency is to err on the side of caution and believe the worst is yet to come. A lot of today's information points in just that direction.

As the Break Point Approaches

Looking at our energy use in terms of a break point provides us with a powerful way of interpreting an otherwise confusing and even random-seeming pattern of events. For instance, what does

President Bush holding hands with Prince Abdullah of Saudi Arabia have to do with Venezuela's Hugo Chavez threatening to sell oil anywhere but the United States and a national Chinese oil company bidding for the independent but American-based oil company Unocal? How does that connect with the conflicting consumer desires for larger and larger SUVs on the one hand, and hybrid-gas-electric-powered economy cars on the other? Does any of that have anything to do with volatility in oil prices, end-of-oil theories, renewed talk about the value of nuclear power, and the announcement that the world's major industrial nations would combine in building a multibillion dollar nuclear fusion reactor in France?

Sometimes a break point is triggered by a seminal event. When whalemen were forced by the scarcity of sperm whales to go to the ends of the earth in search of their catch, they must have known that a break point was near. But whale books often cite two events that can be considered break points. First, in 1861, during the Civil War, a confederate ship set 24 whaleships ablaze, badly depleting the U.S. fleet and driving whale oil prices up at a time when rock oil was starting to emerge as a compelling substitute. Second, in 1871, a terrible freeze-out in the Arctic Ocean trapped 33 ships, sparing none. Though all the sailors ended up being rescued by seven other ships after a harrowing, icy escape, a letter signed by the masters of the 33 ships during the ordeal said: ". . . our ships cannot be got out this year, and there being no harbor that we can get our vessels into, and not having provisions enough to feed our crews to exceed three months, and being in a barren country where there is neither food or fuel to be obtained, we feel ourselves under the painful necessity of abandoning our vessels, and trying to work our way south with our boats, and if possible to get on board of ships that are south of the ice . . ."[10]

Such a note signaled that mankind was at the limits of the ends of the earth in its quest for one of the primary fuels of the day. The break point had been reached, and there was no choice but to switch to other sources.

More often, however, a break point is the accumulation of a collection of events and circumstances that lead to a single, compelling realization: the fuel we are relying on is disadvantaged, not just in a cyclical or seasonal sense but permanently, and hence, we are disadvantaged, and need to make a significant change. It's as though we have been driving along a highway for some time and suddenly recognize that the traffic is not going to get any better or the route we are traveling is no longer going in the right direction. Decisively, we seek out alternatives and get off the highway to take a new road to our destination.

I call this moment of realization the rallying cry. We hear it after a period of pain, complaint, and confusion, and it comes from several sources. In business, the rallying cry occurs when the executives in the board room finally realize that the bottom line is negatively affected to such a degree by the current choice of fuels that some other alternative would be significantly better. When the railway companies finally made their decision to switch from coal to diesel, it was because they had reached a break point in their energy cycle. It was no longer possible to be competitive and stay solvent going forward without a switch. When a variety of industries make similar choices in response to a break point, the impact on the energy supply chain can be tremendous.

But it is from national governments that the primary rallying cry comes, and it is national governments that pack the biggest punch in terms of influencing or even forcing industry and society to make crucial changes. When Winston Churchill decided that the British Navy needed to convert from coal to diesel because coal was disadvantaged as a fuel, that rallying cry signaled a break point that changed the United Kingdom's energy mix. A less dramatic, but no less significant, break point occurred when President Jimmy Carter's administration legislated that Detroit automakers had to make more fuel-efficient vehicles, American car drivers had to reduce their speeds, and American power generators needed to increase their reliance on coal and uranium. This sort of government-mandated and unified effort is typical of what occurs in our modern world when it is necessary to fix a growing energy problem by rebalancing our energy mix.

But the government makes pronouncements about energy all the time, so how do we distinguish background chatter from the rallying cry? As an individual, you will recognize a true rallying cry and the conclusiveness of the break point by how resolutely it affects your life and lifestyle. Driving slower, buying fuel-efficient cars, easing up on the thermostat: These are some of the lifestyle changes that we experienced in the 1970s. What kind of changes will we experience in the next 10 to 20 years? That's the story of the rest of this book.

For now, imagine what it has been like when you have made changes in consumer products in the past. Have you ever used a computer for so long that it was no longer worth it? When you first bought it, it was state-of-the-art, but gradually it lost functionality. Problems occurred, the speed got slower and slower, the memory capacity seemed too small; new connection formats and new software emerged that you couldn't run; then a new generation of operating system was launched that your computer lacked the size to use and you were really left behind. Finally, there came a day when you decided the old computer was too disadvantaged and you bought a new one. Remember when you had a VCR, and the movie-world was your oyster? Along came DVD players and while you were tempted right away, it just didn't seem worth it until one day you walked into the video store and saw nothing but DVDs around you. Remember when your old car became so much bother in terms of maintenance and cost that you knew you needed to buy a new one? Depending on the price of oil, you may have decided that it was okay to get a Hummer, or conversely, that it is better to get a small car that offered great fuel efficiency. If the price of gas gets too high, you may decide to car pool or start taking the bus.

At a break point, those lifestyle changes can seem like painful sacrifices until we readjust. It is the pain of those sacrifices that makes any political administration reluctant to tell the whole truth about our energy situation. Until the evidence of the need for change is more obvious to all citizens, it will be difficult politically to make the necessary tough choices. In the United States and Canada, a sense of energy birthright is deeply entrenched in our mind-sets. We want our energy cheap, clean, secure, and discreet. We want to fill our gas tanks and our furnaces without undue concern, drive long distances without worrying over gas prices, and live comfortably in our temperature-controlled homes, sheltered from the heat or cold outside. We don't want to feel vulnerable to the tensions and conflicts of the Middle East, and we would prefer to reduce or eliminate our reliance on foreign oil. We don't trust big oil companies and don't want them to make exorbitant profits. We fear nuclear power and don't want to see it return to prominence after unforgettable incidents like Three Mile Island in the United States, Chernobyl in Ukraine, and Taiko-Mura in Japan. We treasure clean air and a clean environment, and we don't want to see a return to the heavy use of coal. We certainly don't want unsightly pipelines, refineries, or other energy supply infrastructure anywhere near where we live. We simply want energy available to us, at a cheap price, out of sight, whenever we need it.

Are we prepared to contemplate the trade-offs that will be necessary going forward? Everyone should value and make efforts to preserve the environment—I believe this very strongly. But the technology needed to make the burning of fossil fuels cleaner increases the cost to consumers. Are we willing to accept higher prices for cleaner energy? Our growing energy needs have put the world's energy supply chain infrastructure at maximum capacity, straining our ability to bring energy to the market. Are we willing to accept more pipelines, more refineries, and other intrusive infrastructure development within proximity of where we live, breathe, and work? If not our backyard, then whose backyard should it go in? If we want to reduce our dependence on foreign oil going forward, are we willing to accept an increase in our reliance on clean and relatively safe nuclear power or liquefied natural gas? Do we understand how difficult it would be to switch from gasoline-powered cars to hydrogen-powered ones?

During an era of energy plenty, it is rare that we have to make such tough, painful choices. But when a break point approaches, choices that are painful on the one hand or distasteful on the other are often the only ones available to us. And yet, despite the pain, such wholesale change has historically improved our social conditions and created great opportunities for wealth and economic growth. I guess it depends on how you look at a barrel of oil—half empty, or half full.

Notes

1   *New York Herald Tribune*, January 9, 1948.
2   The national average U.S. retail price of regular, unleaded gasoline that has been inflation adjusted to the end of the second quarter 2005, and quoted in 2005 dollars.
3   For simplicity I have coined the term *oil dependency factor.* Technically, it is calculated in the same manner as the better-known measure in economics called the GDP elasticity of demand and then multiplied by 100.
4   International Energy Agency, *Annual Statistical Supplement 2004 Edition*; Comprised of crude oil, condensates, natural gas liquids, and oil from nonconventional sources.
5   Data taken and extrapolated from U.S. Department of Transportation, Federal Highway Administration, *Highway Statistics 2002*. There were officially 220,932,000 registered light vehicles in 2002.
6   Cadman inauguration speech.
7   Vice President Cheney Salutes the Troops: Remarks by the Vice President to the American Society of News Editors, April 9, 2003.
8   191 members of the United Nations plus Vatican City.
9   OPEC prices relates to an average price representing a basket of production from all the member countries. On average the OPEC basket is cheaper than West Texas Intermediate, because it is a heavier blend.
10 Mrantz, Maxine, *Hawaii's Whaling Days*, page 36.

CHAPTER 5

THE TECHNOLOGY TICKET

"Technology is this nation's ticket to greater energy independence," President George W. Bush declared in an April 2005 speech promoting his Energy Plan. Referring to gasoline costs that were unacceptably high, he went on to tell the expectant crowd, "This problem did not develop overnight."

President Bush was right in his assessment of the situation. Neither the United States' nor the world's energy problems developed overnight; they evolved in conjunction with economic growth and dependency. On the one hand, our problems are evidence of the extent to which our society has grown through crude oil. On the other hand, they are a kind of reckoning for the choices we have made along the way that increased our dependence on a wonderful resource. When we imagine the power of technology to change the world, we often think in terms of stunning speed: a bolt of lightning from the sky, a magic bullet quickly neutralizing all that ails us. But the amount of time from experiment to application is almost always much, much longer than we realize, and in the case of energy technology that often means decades.

Diverted Paths

Take the introduction of the steam engine to the United States, for example. The Boulton-Watt engine was rightly credited with accelerating the Industrial Revolution, however, the revolution began in the United States before the steam engine arrived from England. From the beginning, Great Britain had exploited the United States as a source of raw materials like cotton, while restricting its loom technology by forbidding the export of machinery or even the emigration of anyone who knew how to build it. The knowledge trickled out anyway, and in the early nineteenth century, the manufacture of textiles soon began in New England. Although the transfer of steam engine technology was similarly prohibited, American-made versions were available to the American textile mill owners. Nevertheless, for a time, those mill owners preferred other, more advantageous options already available to them: water power.

The Industrial Power Network was a system of canals built throughout New England that enabled industrialists to transform water energy into the power needed to run their mills. The canals had originally been intended solely for transportation. Some, like the Middlesex Canal in Massachusetts, were quite extensive, covering 27 miles to connect the Merrimack River to the port of Boston via 20 locks and seven aqueducts. The Pawtucket Canal, on the other hand, was only a mile and a half in length but provided a means of traversing the Pawtucket Falls on the Merrimack River at East Chelmsford, Massachusetts, a town that would later be renamed Lowell.

The canals were tapped as a source of water energy when the need for power made the investment in waterwheel technology worthwhile. Well-situated to take advantage of water energy, Lowell became criss-crossed with six miles of canals over the next few decades, transforming its landscape and diverting the path of the mighty Merrimack River. New mills were built alongside these canals, their large waterwheels capturing the flowing water and transferring that energy to the mill through shafts, gears, and turbines. By 1850, the canals of Lowell powered 40 mill buildings, 320,000 spindles, and almost 10,000 looms, providing employment to more than 10,000 workers. By many accounts, the booming town of Lowell was an industrial marvel.

Mills could now operate full-out, all day long, every day of the year. Some even used whale-oil lamps to keep the factories illuminated in the evenings. Yet, the canals were dependent on a strong flow of water and vulnerable to seasonal changes in weather. Accordingly, the mill operators of Lowell did what anyone has ever done with a precious resource: they hoarded it when it was scarce, closing off the canals and leaving the water in ponds. Those living and working downstream suffered the consequences. But even mills upstream were limited, after a certain point, as to how much more capacity they could grow. Fresh, running water churning over the rocks and waterfalls once seemed like an abundant and endless resource in New England, but like other sources of energy, it could be overexploited.

Faced with this break point, the steam engine was waiting in the wings as the magic bullet that would help rebalance the power mix of the region. Despite the effort and capital that had been poured into the extensive and highly regulated system of canals, their utility was simply not comparable to the steam engine. Using waterwheels became a disadvantaged commercial liability. What made the shift from waterwheel to steam engine so rapid was the steam engine's superior utility and the ease with which it could be adopted. The steam engine could turn the same turbine that powered the looms inside the mills, but it took up less space, had a greater power density, and used a more reliable source of fuel that put no cap on future growth. Coal, of course, was not as cheap as water (once the cost of building the canals had been discounted), but it was cheap enough and had other significant advantages like constant supply. Even so, the steam engine itself had been developed, tested, and perfected over the course of many decades and at great cost.

As a species, we are perennial optimists, and the introduction of a device like the steam engine gives us a sense that magic bullets really do exist. I remember such a magic bullet being offered to the oil industry in an earlier stage of my career. On March 23, 1989, a pair of scientists working together announced the discovery of one of the universe's great secrets: They had achieved nuclear fusion at room temperature. Cold fusion had reportedly been achieved by running a current of electricity through "heavy" water. If this were true, it could theoretically provide the world with an almost limitless source of pollution-free power. From this day forward, energy would be cheap, safe, environmentally clean, and no longer vulnerable to geopolitical strife or corporate greed. In essence, cold fusion signaled the end of a way of life as we had known it for the last century and a half.

I'm sure that more than a few oil executives paled at the thought. After all, the 1980s had been a roller coaster ride for the oil industry, making everyone extra sensitive to bad news. In the mid-80s, oil prices were on a distinct downward trend, and acid rain and global warming were emerging as major public concerns, giving rise to a clamor for new technology (such as cold fusion) that would save us from fossil-fuel dependency. The industry's woes deepened in 1988 when the giant oil tanker, Exxon Valdez, ran aground off the pristine coast of Alaska, creating the largest oil spill in history. At the tail end of the decade, the U.S. economy was teetering on the brink of recession, adding to the general feeling of uncertainty.

At the time I was working for Chevron Corporation as a geoscientist, and the news about cold fusion sent a ripple of anxiety and concern through the ranks. Chevron was a storied oil company, one of the original seven sisters that firmly ruled the energy world before the rise of OPEC. As Standard Oil of California, it had been a founding partner in Aramco, a joint venture that had first tapped into the mammoth reserves of light sweet crude discovered in Saudi Arabia. Now its employees were standing around the water cooler, wondering whether they would have jobs in a year. I was pretty sure I would land on my feet, figuring that my math and physics skills would be applicable in some other industry. Others were in less flexible career modes than I was. The concern was noteworthy enough that a memo was soon issued by the executive office in California. I don't remember the exact wording, but the gist of it was a poker-faced message to remain calm. Basically it said, we're studying the news and we'll get back to you.

In fact, everyone was studying the news. Some scientists said that "fusion in a jar" would be the most important scientific discovery since fire or the invention of the wheel. Others were less eager to jump on the bandwagon. The paper that had been written by the two lead scientists, Stanley Pons and Martin Fleischmann, was withdrawn from the esteemed journal *Nature* for suspicious reasons. Some speculated that it didn't meet scientific standards; others said that its contents were so valuable the two researchers would be foolish to divulge the information freely. But early draft copies of the paper were being faxed around madly to the physics labs of the world. In those labs, other scientists were attempting to duplicate the findings, but no one was having success.

Though skepticism in the scientific community was fairly immediate, few were willing to put their cards on the table and declare the experiment a total fraud or a hoax. Maybe that explains why some politicians and media figures were unrestrained in their own enthusiasm. Dan Rather, then the *CBS Evening News* anchor, got the excitement rolling by leading off his nightly segment with the report on cold fusion's discovery. A number of politicians on Capitol Hill quickly followed suit, pledging millions to this important, world-changing research. Although voicing plenty of doubts, *Time* magazine described the stakes nicely: "A practical technique for creating useful energy at low temperatures could change the world forever by providing a source of virtually limitless power. . . . Any scientist who managed to harness fusion would be guaranteed a Nobel Prize for Physics (and probably Peace as well), untold riches from licensing the process and a place in history alongside Einstein and slightly above Edison."[1]

Alas, cold fusion was, as *Time* reported, nothing more than a "fusion illusion." Whether by mistake or intent, the original research presented misleading conclusions, and the success of those experiments could never be convincingly duplicated. Nobody discounted the possibility that cold fusion might happen one day; but for the time being, the world was stuck with oil. The employees and executives of the energy industry went back to work, undoubtedly relieved for their job security. But the hope had been kindled in our collective social consciousness that somehow, through a great leap in technology or innovation, perhaps even by some scientist currently working in obscurity, a magic bullet might one day be offered up that would instantly heal the world's growing energy problems.

No one should hold their breath waiting for our energy panacea to present itself. The development of a radical new energy technology can take decades. To understand that process better, let's consider how Thomas Edison lit up the world.

Bright Ideas

The timely invention of kerosene and the kerosene lamp saved the sperm whale from extinction and helped us rebalance our energy mix. But kerosene couldn't light a city. For that purpose, coal gas was the fuel of choice. First developed in the late 1700s, coal gas, also called "manufactured gas," was refined by mixing coal or shale with a slurry and boiling it at intense heats in sealed vats. The gaseous by-product was cleansed until it yielded a dirty methane. This was then transferred through an infrastructure of pipes under the ground to burners in homes and street lamps. It was almost as though a whale oil lamp were being supplied with constant fuel that never needed refilling.

Making that fuel was a grim, dirty business, producing toxic by-products that were extremely harmful to the environment. Pollution from the burners in homes and streets was considerable, too. Still, despite public health concerns, steep costs, and the opposition of candle makers and lamp oil merchants, the ability to have light on demand from a centralized source made coal gas very appealing. Textile mills in Watertown, Massachusetts, and Pawtucket, Rhode Island, were lit by coal gas as early as 1813, as were the street lamps of Baltimore, Maryland, by 1817. For the next 60 years, more and more city streets, wealthy households, and big factories were lit this way as the number of coal gas manufacturing plants grew rapidly. By the mid-1870s, most large towns in America were serviced by small coal gas companies.

The virtues of coal gas, whale oil, and even kerosene were soon eclipsed by a radically different kind of illuminant: the electric candle. Long before 1878 when Thomas Edison spelled out his vision for an electric light system for lower Manhattan, efforts were being made in Europe and America to develop a lamp powered by electricity. In 1802, for example, Humphry Davey created the first electric arc light when he sent a charge between two slightly separated rods of charcoal. The spark running between those rods was quite bright and could be sustained until the rods burned down like cigarettes and drew apart. For his energy source, Humphry Davey used a voltaic cell battery, invented only two years earlier. Over the next few decades, the arc light was a great focus of experimentation for many scientists. Elaborate and precise mechanisms with gears and levers were developed to press the carbon rods together at a rate that matched their rate of consumption, thus sustaining the light for longer periods of time. All that effort dwindled by 1860, however, as reality set in. Without a better energy source than the expensive battery, there was no practical application for the arc light. Later, their intense illumination would prove too bright for homes and offices anyway, and would find better application in lighthouses and on film sets.

Humphry Davey's energy problem would be resolved by the time Edison and his peers turned to the development of the electric lamp. In 1831, Michael Faraday, Sir Humphry's student and successor at the Royal Academy, discovered the principle of electromagnetic induction. This led him to develop the electric dynamo, a device that transforms mechanical energy into electrical energy. Over the ensuing decades, many versions of this machine were tried by others, and the wondrous practical applications of electric current created a great stir among inventors in various fields. By 1844, for instance, Samuel Morse had invented the telegraph, which transmitted signals "written" in Morse Code over an electric line. Soon, telegraph wires crossed the landscape and became essential to modern life. Thirty years later, in 1877, Alexander Graham Bell tested the first commercial telephone, transmitting the human voice via electric current. Recognizing the compelling need for his new device, Bell envisioned a day when telephone wires would run to houses and buildings like coal gas pipes did.

The rapid commercial success of the telephone had a significant impact on young Thomas Edison. Financially, it made his early fortune. Edison had invented a carbon button that was essential for the microphone in Bell's telephone, and orders for it kept his famous laboratory in Menlo Park, New Jersey, humming. But just as important as the financial boon Edison received from his contribution to the telephone, he was also influenced by its path to success. Edison recognized that an invention was pointless if it did not meet a significant need. He had already learned that lesson in his early twenties when he went broke investing his savings and talent into developing a vote-counter, a device that actually worked but could find no buyer. He vowed never again to devote himself to anything that did not have strong commercial potential. As a result, his later inventions, like the phonograph and stock ticker, were developed with built-in commercial markets in mind. When Edison encountered the idea of electric light, he understood as well as anyone that this would be a revolutionary and profitable endeavor.

According to contemporary newspaper reports, Edison first latched onto the idea of electricity as a power source when traveling out west in the Rocky Mountains. Observing the hard work done by miners to drill into rock, Edison wondered why the energy of the strong coursing water in nearby streams couldn't be transmitted via electricity to power tools at the mines. Back in New York City, Edison continued to ponder the significance of electricity. Soon, he was shown an electric dynamo, which the acquaintance who built it called a "telemachon." Upon seeing its capabilities demonstrated, Edison was immediately seized by a vision of how electricity could be used to distribute light. In a burst of activity, an experiment was soon set up. When power was drawn from the telemachon, Edison was able to momentarily illuminate eight electric lights at the same time—equivalent in intensity to 4,000 candles, he suggested. Within a week, according to those exciting newspaper reports, Edison was convinced that he had discovered how to make electricity a cheap and practical substitute for coal gas. Where others had tried and failed, Edison had triumphed. And he had done so by a path that none of them had ever even considered. Although the solutions had come to him easily to this point, he cautioned that much hard work would be needed to perfect the technology. Nevertheless, he foresaw that he would soon be able to illuminate lower Manhattan through the power of electricity.

How would he do it? Edison saw the network of gas pipelines not just as a metaphor for how to construct an electric power grid, but as an existing system that could be piggybacked to distribute electricity. Insulated wires would be run through the gas pipes. The gas burners and chandeliers already in factories would serve as light sockets, making the "electric candle" easy to adopt. Meters could be installed to measure the use of electricity, and customers could throw their gas meters and matches away. The same electricity could also be used to heat homes and power any mechanical devices that used a motor. Moreover, all of this could be done at a fraction of the current cost. The work of three dollars of coal gas, Edison calculated, might cost as little as twelve cents using electricity. It was only left for Edison to get to work and build his system. He knew that he was entering the game somewhat late, but he was confident about his prospects for success. "I have let the other inventors get the start of me in this matter somewhat, because I have not given much attention to electric lights; but I believe I can catch up to them now,"[2] he said in 1878. There were a few small technical challenges ahead of him, but no matter, every puzzle would be solved in short order.

It's a good story, and a colorful depiction of the man that contemporary admirers, enthusiastic reporters, and a bedazzled public would call "The Wizard of Menlo Park." In fact, Edison's lifetime accomplishments were monumental but not necessarily as we have been taught in school. Wealthy in his twenties, he took out over 1,000 patents in his own name and saw his legend as a scientific capitalist set in his lifetime. Although we remember him today as the genius who ultimately won the race to produce a working electrical grid system, the truth is somewhat more complex. An examination of the history of patents associated with the evolution of electric light shows that Thomas Edison was not only late to the game, but late to the realizations that he declared to be so astonishing and original that no other scientist had even considered them. In fact, the patent record reveals that once Edison did get into the game, his ideas for how to solve the many so-called minor challenges ahead of him were wrong more often than they were right, and he would end up backtracking and following the paths of those he derided publicly in the battle for electric supremacy.

So how did Edison end up ultimately winning that race and setting many of the standards that we would live by in our bright electric future? Edison was a master at envisioning how technology would impact our lives and what kind of business organization is required to nurture and support that technology on the road to commercial viability. In that sense, he was the Bill Gates of his day (or Bill Gates is the Edison of his day), a man of formidable creative intelligence in his own right but who owed some of his market success to his ability to capitalize on the accomplishments of others in a very competitive field. If Edison's famous adage that, "Genius is one percent inspiration, ninety nine percent perspiration,"[3] could be expanded on, it might go something like this: Edison realized that dreaming up and perfecting an invention was not enough to establish its adoption. The successful inventor also needed to convince financiers to bet on his idea, convince the media to amplify enthusiasm for it, convince those in political power to see a need for it, and convince customers to clamor for it. He saw electric light, not as a bright lamp isolated unto itself, but as a final destination at the end of an electricity supply chain in the context of the marketplace.

As a way of highlighting the qualities that Edison brought to the game, it's worth comparing the path taken by one of his most important competitors, William Sawyer. Like Edison, Sawyer had been a telegraph operator in New England. Later, he became a reporter in Washington, DC, before becoming obsessed by the problems of electric light. In a patent filed in the summer of 1877, one year before Edison jumped into the game, he described his plan not just for electric light but an entire electric system that would bring power to city blocks, buildings, and houses for light, heat, and motors. That system would function in the same way that gas and water is supplied, using the infrastructure of pipes as a channel for the wires and the gas burners as light sockets to make power available by opening a "stop-cock" and allowing the desired power to flow through as needed.

Sawyer even described the meter system that would be used to measure and charge for the amount of electricity being used by the customer. In his next patent, filed a week later, he mapped out the engineering required for using a single generator to power many lights in parallel. This was the subdivision of electricity that Edison's newspaper chronicler would describe a year later as "being a thing unknown to science." But Sawyer, in his patent, would describe it as an old idea, for which he was not claiming credit. It had been worked on by many before him, at least since 1870.

Sawyer's efforts toward his vision were proceeding when Edison came on the scene. No doubt Edison (as opposed to the enthusiastic reporters who wrote about him) was aware of that progress and the challenges still to be met. Most of those challenges were technical, but many of them were not. Financial backing was crucial, of course. In this area, Sawyer was hampered by a lack of resources, so he teamed up with a business man from Brooklyn named Albon Man who became equally passionate about the endeavor. Edison, on the other hand, parlayed his credibility and star power—what we might today call his brand—to gather the support of a group of investors who included the great financier J.P. Morgan. Together, they formed Edison Electric Light, the mother company of General Electric.

As for technical challenges, one of the first and most significant was to discover the right substance to use as a filament to create illumination when charged with electricity. Edison first tried carbonized paper, but without success. He declared carbon to be a dead end, and turned to platinum. Sawyer and Man, on the other hand, had abandoned platinum the year before in 1877, and were focusing on carbon. After a year of working with platinum, Edison had still not come up with a viable solution. As the story goes, he was rolling a cotton thread between his finger and thumb, blackened with tar, when he had an epiphany that a thin carbonated thread would do the job. This was described in newspaper headlines as "The Great Inventor's Triumph In Electric Illumination."[4] In fact, Edison's application for a carbonized filament patent would be finally rejected after five years of legal battles, because of a previous patent awarded to Sawyer Man for the same idea. This forced Edison to file his patent for the electric light bulb using a filament made from a strip of bamboo. The patent was granted, but the bamboo filament never made it into final production.

There were many other challenges of course. For instance, Edison believed that the amount of electricity should be regulated at the lamp. To that end, he worked diligently to come up with a system that would allow consumers to adjust the brightness of their lamps. Sawyer had long proposed that power, like gas flow, should be regulated at the central generator, an idea that Edison eventually adopted himself.

Given Sawyer's superior technical achievements and instincts, why is it that we remember Edison as the father of electric lighting and not Sawyer? In part, this can be blamed on Sawyer's poor business skills. Though an engineering genius, he was a suspicious man, difficult to work with and wasteful of money, who apparently drank heavily. He was later forced out of his own company, the Electro-Dynamic Light Company of New York, formed a rival, and went to great lengths to try to discredit his first enterprise. More often than not, Edison outmaneuvered Sawyer legally on patent applications, and in the eyes of the press, public, and investment community, Edison was the bankable star. To put it simply, Edison knew how to create buzz and he was adept at using the media to further his enterprise. The news of his astounding technical advances was carefully crafted and released to admiring reporters who wasted no time getting the story in print, often without fact checking or questioning other sources. As a contemporary observed, "The most absurd and exaggerated statements were eagerly printed by the newspapers and still more eagerly swallowed by an insatiable public."[5] So well did he control the media that Edison's pronouncements of his own success with electricity often caused shares in coal gas companies to plummet. Like euphoric Internet stock watchers 100 years later, people were so convinced by Edison's vision for his wonderful technology, they expected the world to change overnight.

Edison knew how to make that changeover something to be desired. One simple but important example comes in his choices for a lightbulb base. In coming up with the best design, Edison's primary criterion was to develop a base that would screw into any lamp burner currently on the coal gas system. He knew how important it would be for consumers to find the switch from coal gas to electricity as simple as removing a burner and screwing in a lightbulb.

The first public demonstration of commercial electric illumination occurred in 1880 when Edison lit a giant steamer, the Columbia, which dazzled the people of Manhattan watching from the shore at night and later set sail for San Francisco. The first house he lit was J.P. Morgan's, whose neighbors complained about the noise of the generator. The first commercial generator he built was on Pearl Street in 1882 to light lower Manhattan, an area that encompassed the heart of the financial district. What better way to secure the enthusiasm of the investment community than to power the lamps on their desks? It was probably not overlooked by Edison that in lighting up those bankers' offices, he also lit up another building in the neighborhood, one that was home to the *New York Times*.

Not even Edison won every battle over standards, however. When a significantly more compelling option was available, the market took to it. For instance, Edison's electric grid for lower Manhattan was based on a DC or direct current system. This approach worked, but had a crucial weakness: it required a power plant to be built within close vicinity of the customer because DC current weakened over short distances. As a result, even a geographically small but densely populated area like Manhattan would require hundreds of power plants to run DC-powered electrical devices. A much more practical system used AC or alternating current, an approach invented by Nikola Tesla and adopted by George Westinghouse, whose company was in competition with Edison Electric. Like the candle makers and lamp oil merchants who resisted the introduction of coal gas, Edison claimed that alternating current was dangerous and even deadly. To demonstrate, his supporters performed public electrocutions of many small animals, and one large one, Topsy, the Coney Island elephant. But Tesla's AC approach made a lot more sense economically, and it was adopted as the industry standard over Edison's strong objections. Interestingly, vestiges of the DC system remain in existence today in parts of Manhattan. After decades of trying to encourage the last 1,600 customers to switch to AC, the local electric company, Con Edison, finally put out notice that DC power would be shut off for good in 2005, 123 years after it was first turned on.

In the early history of the electric industry, many such battles took place. There was an understanding that the competition was important, since the winner would be leading us into the future. The development of electric lighting seemed so linked to the notions of scientific and social progress that the public was engrossed in the evolving story, following the attacks and rebuttals between rival companies and systems in the newspapers. The only interested parties that didn't seem to pay attention to these battles were the coal gas companies, which ignored or failed to embrace the new lighting system and would see their market share essentially disappear within a few decades.

The Last Carbon Standing

Once compelling standards are set they are remarkably intransigent. This fact alone makes the period from 1800 to 1920 the most influential in our modern energy era. Despite all the scientific progress we have achieved in the interim, in our search for energy alternatives we remain largely fixed in our approaches and limited in our options because of the winning technologies that were adopted in those early, formative days. Think, for example, how difficult it would be to successfully mass market and sell a 60-Watt lightbulb that didn't fit Edison's simple screw-based lamp socket, or an electric product in North America that didn't run off the two-prong 110-volt wall plug. Consider, in an entirely different field, how odd it is that today's keyboards still use the QWERTY key standard. It was developed for the typewriter in 1872 to prevent the hammers of frequently used letters from jamming, but while the two Gigahertz dual processor computer in our office has no such worries, that doesn't mean we can easily adopt a more efficient approach. The same goes for the operating system on that computer. Microsoft Windows may not fulfill our dream of how a PC should operate, but as rival systems have discovered, it remains incredibly difficult to dislodge a standard once it has become mainstream.

And so, too, should innovators and pundits in the world of energy consider that proposing alternatives is not just about displacing an obnoxious fuel or a dirty engine. Rather, introducing an alternative is about replacing an entrenched set of compelling standards up and down a complex supply chain. Gasoline, for example, is not merely a fuel, it is the Microsoft Windows operating system of the transportation world, and much harder to displace. It is in this light that Edison's skills as an innovator can be best appreciated. In fact, his relentless determination to win support for the electric lightbulb was absolutely necessary to overcome the deeply entrenched standards of the existing energy supply chains—for example, kerosene, coal gas, candles, and even the last vestiges of whale oil. Imagine switching a whole world to a new source of illumination. Anyone who longs to create such change today will need all of the skills of Edison, and then some.

Nevertheless, there are those today who believe that new alternatives to oil will come to the rescue anytime now. Even the venerable Alan Greenspan sells a version of this story with his public retelling about how "oil displaced coal despite still vast untapped reserves of coal, and coal displaced wood without denuding our forest land."

In fact, Mr. Greenspan's belief in the relatively calm transition from one primary energy source to another is not uncommon. As an equity analyst following alternative energy stocks in 2000, the idea that oil displaced coal without the world running out of coal was a management mantra for companies involved in high-tech alternative energy solutions. It was especially commonplace for companies involved in developing the "hydrogen economy" to proclaim that the end of the oil age was nigh.

On the surface such beliefs are not wrong, but it's dangerous to think it will all just work out and a solution will emerge before we run out of a primary fuel like oil. History tells us that this simply isn't so.

Figure 5.1 shows the history of the U.S. energy mix going back to 1645. Using data from the U.S. Department of Energy, the mix is represented as fractional market share. So, in 1800 the nation's exclusive energy source was wood. After railroads were introduced in 1825, and textile mills started converting from the water

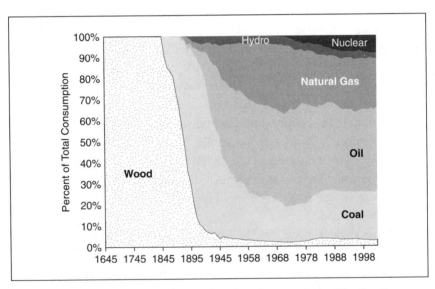

Figure 5.1   Long-Term Evolution of the U.S. Energy Mix: By Percent of Total Consumption (*Source: Adapted from U.S. Energy Information Agency and ARC Financial*)

canals of the Industrial Power Network to coal-fired steam engines, coal began rapidly displacing wood. Oil followed suit around 1859. Natural gas made its debut shortly thereafter, as did hydroelectric power in around 1880. Nuclear power, the little wedge in the upper right corner, made its commercial debut in 1957.

Of course, there is a major omission in the Department of Energy data since whale oil was a major source of lighting energy between 1770 and 1875, and should be sandwiched at the top, just before the oil age. I don't have a good sense of what share of the U.S. energy mix was comprised of whale oil, but my best guess is that it would be around 10 percent. If we were just considering the market for lighting only, that percentage would be far greater.

The numbers are really not that important; the real point is that there is precedent for fuels almost running out before a new fuel is adopted. The world's whale population was effectively "denuded" by the 1870s. By the end of the era, tales abound of whalemen floating for weeks on end without sighting a single one of the creatures. Certainly, whales were completely absent in the North Atlantic, and most everywhere else, which is why whaling ships eventually had to travel all the way to Alaska to find them.

In comparison, the progressions from wood-to-coal and from coal-to-oil were relatively unstressful. One hundred and fifty years ago it really wasn't difficult to find these commodities. Trees were all around, coal could be found at surface, and rock oil seepages would point people where to drill. The forces that catalyzed growth in their usage were admittedly more complex—the steam engine for coal, and the internal combustion engine for oil. But all in all, the progression was not that difficult a leap to make, though in the United Kingdom, a nation that was more advanced in terms of energy use than the United States at the time, forests were almost denuded before coal arrived on the scene.

The next transition was to natural gas. Again this transition was not too challenging to make because natural gas is almost always present when you drill for oil. Hydroelectric power was even more of a no-brainer. Industrial canal networks proved that rushing rivers were a natural source of power. With the refinement the dynamo, invented by Michael Faraday in 1832, we were able to convert the energy from a spinning magnet into electricity. Adapting the device to a rotating waterwheel was a natural application that provided the basis for hydroelectric power even to our own day.

The latest addition to our energy mix—nuclear power—required tremendous human resources. Though uranium is found naturally like coal, oil, and natural gas, its energy capabilities were not obvious to anyone. Figuring out how to tame and unleash the power within uranium required the best brains in the world in 1936, when the Manhattan Project was launched at the orders of the president. The first atomic bomb was dropped in August, 1945. Twelve years later in 1957, the first commercial nuclear reactor went into service in the United States. While wood, coal, oil, and natural gas were there for the taking, uranium took time and substantial technological prowess to turn into useable electricity.

With time at a premium, how long should we expect to wait for the next great energy substitution? To date, the substitution rate for wood was actually the fastest in energy history, but even then going from wood to coal took 75 years. Oil took the bulk of coal's market share away between 1860 and 1960, a substitution that took 100 years. Natural gas has never taken more than about 30 percent share, but its rate of substitution was even slower than coal's. Historically, substitutions in the energy world take a long time and there's no reason to think the next substitution is going to happen overnight. While it's likely that we may not run out of oil before a substitute is found, it will be decades or more into the future before any new solutions make a difference.

Why was the substitution of coal over wood the fastest switch we've seen in energy? First and foremost coal is a much more compelling fuel than wood. It packs more energy per unit volume, burns hotter, and doesn't rot. Less obviously, but more important, coal didn't have to overcome a large infrastructure of supply chains entrenched with standards. Edison's success in the lighting industry showed that adoption of new energy supply chains is not all about better technology. Edison made it happen only after understanding and overcoming barriers of incumbent standards and competition. Sometimes those competitive forces are formidable. Candle makers declaimed whale oil as a fuel. Edison fought against alternating current. When electric refrigeration emerged as a technology that could displace the ice harvest in the early twentieth century, the ice harvest industry lobbied furiously to have refrigeration banned with claims that it would poison food.

Some people mistakenly compare the rate of change in the energy world to that of the high-tech world. One of the most comprehensive substitutions today can be seen in how DVDs are quickly making videotapes obsolete. The first commercial DVDs really started emerging in 1997. I think it's safe to say that by 2010 finding a videotape is going to be about as easy as finding an 8-Track tape. The entire base of videotapes will be substituted out in just over a decade, and by that time digital video technology will have been embraced in many other markets as well, from its early origins in geophysical imaging to digital cameras, and television and telephone signals. Even whale oil lamps took longer than that to be made obsolete. There is no precedent in the history of energy—either fuels or devices—where substitutions happened as fast as that in consumer electronics. The fastest substitutions of energy devices were probably diesel engines over steam engines, and jet engines over propeller engines, each of which took over 35 years. Comparing energy substitution rates to those in today's high-tech arena is like comparing apples to oranges.

Nor is it a good idea to think that the rate of radical innovation is the same in energy as it is in the world of consumer electronics. Witness the number of major step-changes in recorded sound over the past 100 years. Edison's recording cylinder was invented at the turn of the last century, followed by the 78-rpm bakelite record. Next came singles played at 45 rpm and then long-playing records played at 33 rpm. Then the reel-to-reel tape, the 8-Track, the cassette, the CD, the MP3, and the many audio files that played on the iPod. That's at least 10 radically different technological platforms that have arisen in the sound industry, three of which have been in the past 20 years.

Conversely, the pace of radical change in the energy industry is slowing, not speeding up. Consider how many new large-scale energy sources (platforms) have been introduced over the past 100 years. The answer is one: nuclear power, which was introduced almost 50 years ago in 1957. How many have there been in the entire history of energy consumption? Eight, if you add whale oil and animal fat (candles) to wood, coal, oil, natural gas, water, and uranium. A handful of renewable energy sources can be included on the list, too, such as solar, wind, geothermal, and wave power. However, their lack of scalability makes it very difficult for these energy sources to make fast, large-scale contributions to an industrialized nation's overall energy mix. They can help rebalance after a break point, but none of these alternative sources has the scale or the flexibility to stave off a break point when it comes to a vast energy source like crude oil. Finding the next major energy supply chain that is based on a completely new fuel source is much more difficult today than in times past.

Of course, there are many clever scientists working on finding the next radically new energy supply chain, experimenting with things like nuclear fusion generators where the raw input is nothing more than purified sea water. A contained fusion reaction is man's attempt at creating a miniature sun here on Earth, the hot version of cold fusion. Research has been progressing on this promising technology since the 1960s, but the technological challenges of containing a fireball that is several million degrees hot means that large-scale commercial use is still a few decades away. And even after it's commercially introduced, fusion power will go through a lengthy adoption period before becoming a significant part of our energy mix.

So what fuel is left to exploit? Today, we have scoured the earth for energy sources, from pole-to-pole, from the bottom of the ocean to the tops of the mountains, from plants to animals, from solids to liquids to gases. The fact is, we've run out of carbon-based alternatives. Wood is composed of complex organic living matter like cellulose and lignite, which is very rich in carbon. Coal, as previously discussed, is essentially decayed plant matter and is also rich in carbon. Rock oils, especially lighter grades, have far less carbon content than coal. Finally, natural gas is principally composed of methane. As the simplest hydrocarbon, a methane molecule contains only one carbon atom surrounded by four hydrogen atoms. Over the past 200 years, we've experienced a great progression of substitutions from wood all the way up to natural gas, because less carbon content means fewer harmful emissions. That's why natural gas is environmentally preferable to coal and oil. But natural gas is the end of the carbon lineage; it's the last carbon standing.

The Most Abundant Element in the Universe

There are a handful of other radioactive energy sources like uranium. But all are less abundant in nature, and more toxic to boot. Putting those aside, the periodic table of the elements—the 100 or so atomic building blocks of the universe—has been exhausted for potential new sources of primary energy, with the exception of one element: hydrogen.

Many are betting that hydrogen will be the wonder fuel of the future. The first and simplest atom on the periodic table of the elements, hydrogen is an odorless, colorless gas. When burned, it combines with oxygen, creating heat and water as a by-product. Burning is not necessarily the best way of harnessing energy from hydrogen because it's a pretty explosive reaction. A more high-tech solution is to use something called a fuel cell, a device akin to an atomic cheese grater that strips electrons off hydrogen atoms to generate electricity. After all, electricity is nothing more than a stream of electrons. Though it's not a combustion process, oxygen from the air is essential to a fuel cell's operation. Again, heat and water are produced as by-products.

That sounds like a dream process: hydrogen and air being fed into a device with no moving parts; water, electricity, and heat come out. Why aren't we rushing to use this process? We're trying to. Governments around the world are funding hydrogen and fuel cell research. Giant automakers like Toyota and GM are pioneers, and have devoted hundreds of engineers to make a practical and economical vehicle that runs on a fuel cell "engine." It all sounds great, but there is a core issue that needs to be resolved: where will the hydrogen come from?

During the peak of the tech bubble in 2001, grandiose press releases from some alternative energy companies made statements like; "Hydrogen is the most abundant element in the universe." The SEC can't dispute this statement, because it is full, true, and plain disclosure, but virtually all of the hydrogen in the universe is contained in the stars, still slightly out of our reach. Back home on planet Earth, the next largest accumulation of hydrogen is in our oceans, lakes, and rivers, combined with oxygen in the form of water.

The water molecule, H_2O, contains two hydrogen atoms and one oxygen. The water molecule is very stable and stubbornly strong and requires a lot of electricity to break, in a process called electrolysis. Of course we know how to generate electricity, but that gets us back to our existing energy mix for electrical power, which in the United States is 51 percent coal, 3 percent oil, 16 percent natural gas, 7 percent hydroelectric, 3 percent renewables, and 20 percent nuclear. In other words, our quest to make hydrogen a new wonder fuel brings us full circle back to our mix of existing fuels. In fact, despite all the industry pronouncements, hydrogen does not exist by itself on Earth. In order to liberate it, energy that we're trying to avoid using must be expended before hydrogen can generate electricity. In other words, hydrogen is not a new energy source at all, but actually an intermediary energy carrier that slots nicely into our existing supply chains.

Although it's not a new source of naturally abundant energy, hydrogen does have distinct advantages that make it well worth pursuing. Cleanliness, of course, is paramount. The lack of moving parts in a fuel cell engine, as opposed to the myriad of gears, pistons, cams, and bearings in an internal combustion engine, make it lighter and more energy efficient, because losses to frictional heat are minimal. Importantly, fuel cells are not bound by the same physical limits of efficiency that inhibit internal combustion engines. Theoretically a fuel cell can exceed 80 percent efficiency, though practically that limit is unlikely to be achieved.

Nearly every automaker has one or more prototypical fuel cell vehicles on the road today. I have personally test driven one and can tell you that these vehicles are marvels of innovation, but the challenges to bringing such vehicles to market are greater than any Edison encountered when pioneering his lightbulb.

First of all, it's worth pointing out that a fuel cell vehicle isn't the first attempt by the auto industry to introduce a radical new engine into the market. Back in the late 1950s, hundreds of engineers were busy fooling with turbine engines, the same type of engines that had taken the airplane industry from propellers to jets. This article from the March 1958 issue of *Popular Science* says it all: "All the noise about small cars drowns out any talk about turbines—the much-discussed "engine of tomorrow" only a few years ago. But Chrysler continues to keep over 100 engineers on the project. There must be a future in it; that's a lot of money to invest. Advocates say turbines will be for sale in seven years. Problems of economy and metallurgy have been licked. Even the durability angle is solved. Of course, piston-engine men quietly laugh at such predictions. And so the battle goes. One group minimizes the gains of the other in the big battle for budgets."

It sounded like all the problems were solved, but turbine cars never did make it to the auto market. In the end, the overall utility of a turbine engine in a car could not compete with its piston-firing competitor. Simple, though ultimately insurmountable, problems got in the way. Turbines take time to "fire up," whereas a gasoline engine is on pretty much instantly. Noise and poor fuel economy were big issues as well. And don't forget that internal combustion engines were being improved upon greatly during that time period, so the new entrant had to compete against a moving target that was getting better and better all the time. Simply put, the utility of a turbine-powered car was never compelling enough to foster market appeal. In the end the piston-engine people had the last laugh.

Despite clean and simple operation, fuel cell vehicles today do not have an overall utility advantage over piston-fired cars, trucks, and SUVs. That's not to say they won't get there eventually, but best-case scenarios appear to be 10 to 20 years from now. Remember too that cost is not the only competitive factor. People will want assurances that a fuel cell vehicle is just as safe, just as easy to start up, just as reliable, just as easy and convenient to fill up, and has the same range and convenience as the gasoline-powered car or SUV. But having the same utility as a regular vehicle today still won't be good enough. If a DVD player was only as good as a VCR, what would be the incentive to switch? As it turns out a DVD player has compelling utility over a VCR—no need to rewind, direct access to segments, far superior clarity, and so on. Think back to the last major substitution in transportation, the diesel locomotive over the steam locomotive, and recall the compelling utility of diesel over coal/steam. If railroads didn't switch to diesel, they went bankrupt. Will anyone go bankrupt if they don't buy a fuel cell vehicle? As a consumer choice, it's going to have to be really compelling—or legislated.

And then we come back to the issue of hydrogen. Set aside the nontrivial problem of having to establish a nationwide network of new fueling stations while mothballing gas stations; where is the hydrogen going to come from practically? In 2005, President Bush pushed for more nuclear power plants to be built, in part because the electricity can be used to electrolyze water and produce hydrogen for fuel cell vehicles in the future. If that's the intent, then more power plants are certainly required. Today's nuclear reactor fleet doesn't have much spare capacity to start fueling vehicles on top of the rest of the nation's electrical needs. In fact, to completely substitute the oil-to-gasoline supply chain with one that goes from uranium to hydrogen—in other words, to completely switch over 230 million cars from gasoline to hydrogen—would require about 350 new nuclear power plants by my calculations. It's going to be tough enough to license three proposed nuclear reactors in the United States and get them by the environmental and NIMBY lobbies, let alone another 350. To accomplish the same feat with clean coal, you'd need over 1,000 new electricity plants commissioned. So, hydrogen is not a free lunch. Coal prices and uranium prices have already doubled in the past two years due to increasing worldwide demand. Today's transportation burden on the energy mix is immense. Shifting the burden to a nonoil supply chain in the energy mix is neither straightforward, nor without consequence to the other fuel sources.

Obviously, to ease our current oil demand problems we don't need to go to the extremes of replacing the entire fleet of gasoline vehicles with hydrogen. Partial solutions are possible, though even so, the magnitude of displacing the oil-based supply chains with hydrogen is staggering and would take decades to implement.

New Routes on Old Roads

So if hydrogen is not the answer in the next 10 to 20 years, where will we find the solution? In the broadest sense, the energy world is like a map with many freeways, highways, roads, and trails. Multiple energy routes may be taken to accomplish the same work. As we've seen, the bulk of the energy feedstock for electrolyzed hydrogen really traces a path back to carbon-based feedstocks, not water. Aside from gasoline, crude oil may also be refined into diesel fuel that can be combusted in an engine to turn a set of wheels. Combusting natural gas in a spark-ignited engine, or extracting hydrogen from natural gas, are yet two other energy pathways that lead to a set of wheels turning. Figure 5.2 shows these possibilities which represent only a minor subset of all the energy "pathways" in our society today.

Human nature is to take the path of least resistance wherever possible. In the energy world, this means that society will migrate to energy pathways that offer the greatest utility at the lowest cost for the work that needs to be done. There is a direct correlation between energy efficiency and cost efficiency. Just as nobody will knowingly take a new route that lengthens a journey, society is unlikely to adopt new energy pathways that are less efficient than

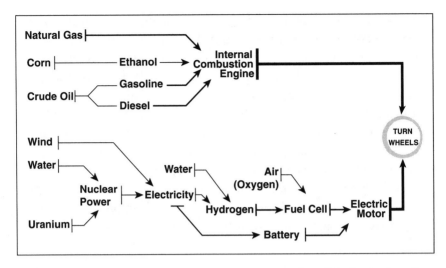

Figure 5.2   Example Energy Road Map: Many Possible Paths Leading to the Same "Destination"

what is already available. Furthermore, the capital cost of "road construction" must be taken into account also. Ultimately it will be society, whether through government subsidy or corporate investment, that will have to finance the opening of any new energy pathways. Those pathways will have to be cost efficient, in other words, competitive with incumbent routes, for capital spending to occur.

Many hydrogen-based pathways are being proposed today, and extensive research and development efforts are underway to optimize the efficiencies along each of the "route segments." Multinational oil and gas companies like Shell, BPAmoco, and ExxonMobil are active on the upstream end of the pathways. Other companies like Ballard Power, United Technologies, and most of the major automakers are active further down the supply chain. A truly united "construction" effort will be required to turn hydrogen pathways into economically viable energy freeways. Many paths will never be paved, especially if the capital cost of construction is amortized into the projects.

But the global energy map is very large. New paths based on hydrogen represent only a subset of the many possibilities for more efficient pathways. In cars, as we've mentioned, gasoline engines, new-age batteries, and electric motors are being combined in hybrid-electric configurations to enhance efficiency. Diesel engines are also being similarly configured.

Battery-driven electric vehicles have long been a legitimate short cut from an efficiency perspective, but have never caught on, principally because of range limitations. Having to recharge a car every 50 to 100 miles is not appealing to drivers long-conditioned to getting 400 miles on a tank of gas. The battery-electric energy pathway seems like a great shortcut, but is too much of a goat path to entice highway drivers to take the detour. And besides, the electrical grid today strains enough under the demand for electricity, how will it cope with everyone recharging their cars? As in any economic problem, it is utility that must be optimized, not just cost. But gains in energy efficiency are not limited to advances in energy technology. For instance, the use of lighter weight materials or more aerodynamic designs are indirect ways of improving transportation efficiency. Such factors add more dimensions to the problem of predicting the most optimum energy pathways.

Another obvious solution is to improve the existing pathways that get energy from the "well to the wheels." Some studies suggest, for example, that the *Well to Wheels Supply Chain* illustrated in Chapter 4 (Figure 4.8) can be improved up to 50 percent over the next 20 years. In other words, well-to-wheels efficiency could rise to over 20 percent through improvements to all conversions above the supply chain.

If significant improvements in current pathways are really coming, then new-age energy shortcuts are competing against a moving, improving target. This is important when assessing the adoption of new energy pathways. Consider that the economics of candles, whale oil, and coal gas were fairly static, and it still took electricity 30 years to gain any serious market share! Of course, the not-so-wild wild card is environmental legislation, which would, so to speak, force us into taking alternate energy pathways by closing down older ones. But if the gasoline pathway can be resurfaced before reaching the political barricades, then the path of least resistance dictates that the alternative routes will remain unexplored.

Technology can still do a lot to improve the efficiency of society's energy map, whether by constructing new pathways or improving old ones. There is no doubt in my mind that we as a society can decrease hydrocarbon consumption significantly without compromising economic productivity. In my opinion, this "alternative" notion is the true potentially radical idea that new energy technology companies need to pursue.

We need to be honest with ourselves. A magic technology bullet, similar to ones that have "saved the day" so many times in the Energy Evolution Cycle (Figure 1.1), will not be in our hands any time soon, and certainly not in time to prevent the upcoming oil break point. Recognizing that, we are going to have to find new ways to deal within the confines of our existing web of large-scale supply chains rooted in coal, oil, natural gas, hydro power, and uranium. This limiting situation is directly proven out in the price of all these primary energy resources, which have all been rising over the past several years. Indexed to January 1, 1999, Figure 5.3 charts

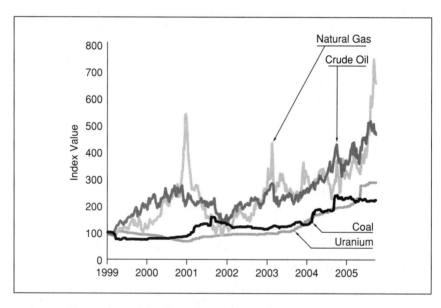

Figure 5.3   Indexed Primary Energy Resource Prices: January 1, 1999 = 100 (*Source: Adapted from Bloomberg and The Ux Consulting Company. Note: Crude Oil is "WTI," Natural Gas "Henry Hub," Coal "Pennsylvania Rail Car," Uranium "U308"*)

each one. Oil and natural gas prices are up fivefold; uranium is up threefold, and coal is up over twofold. With no external relief at hand, pressure indicators are rising within the entire energy complex. It's especially acute in North America, where continental natural gas prices have been appreciating in tandem with crude oil.

Collectively, smaller-scale supply chains emanating from renewables will contribute to solutions as well, but ultimately we must be determined to make all these existing supply chains more efficient, while also making them cleaner and learning how to lower our demand for them. It's a big repaving effort on all our energy highways and side streets, and just like road construction, it won't come cheaply.

If that recipe of solutions doesn't sound very new, radical, or sexy, you should recognize that it's rare in the history of technology that the early adopter gets the worm. More often than not, companies, nations, and individuals who benefit from technological transitions are the ones who learn to take full advantage of the opportunities that are currently available. As I aim to show you in the next chapter, those opportunities are significant. Technology is the ticket, but you may be surprised where the ride is going.

Notes

1 *Time* magazine, May 8, 1989 "Fusion Illusion?"
2 *Evolution of the Electric Incandescent Lamp* by Franklin Leonard Pope, p. 15; 1894, Boschen & Wefer, New York.
3 *Harper's Monthly*, 1932.
4 *Evolution of the Electric Incandescent Lamp* by Franklin Leonard Pope, p. 38; 1894, Boschen & Wefer, New York.
5 *Evolution of the Electric Incandescent Lamp* by Franklin Leonard Pope, p. 18; 1894, Boschen & Wefer, New York.

CHAPTER 6

THE NEXT GREAT
REBALANCING ACT

We are approaching another moment in the evolutionary cycle of energy supply and demand where the *status quo* will be shaken. A break point is coming before the end of this decade. This will force nations to alter the structure of their energy supplies and fuel consumption, especially oil. In the interim, governments, corporations, and individuals need to make choices that minimize the economic damage that can result from the pressure buildup in society's vital energy supply chains.

Light, sweet crude oil is quickly becoming "disadvantaged" as a fuel. Rising energy prices are reducing people's disposable income, eroding corporate profitability, and making a broad range of previously marginal substitutes like bitumen attractive. But break points are not just about numbers. Security of supply and concentration risk for this most strategic of energy commodities is at the fore of political and economic issues. Independent oil companies and their state-owned foreign rivals are embarking upon another great scramble for the world's remaining oil concessions. And like the final years of the whaling industry, the scramble is happening at the ends of the earth, in some of the harshest geographies and climates, the deepest oceans, or the murkiest political regions.

How will the world deal with this coming break point? The past can provide us some answers. In the mid-1980s, the world came out of a difficult and painful 13-year break point and rebalancing period. In many ways, we were all better off as a result. By 1986,

U.S. oil consumption was far less sensitive to economic activity as compared to the years leading up to 1973; only half as much oil was needed to fuel each new dollar of GDP growth. The world had expanded its portfolio of primary fuels in its energy mix to include more nuclear power and natural gas. In countries like the United Kingdom and Japan, oil consumption flattened even while economic growth continued.

Today, each nation is unique in terms of its energy mix and dependency. Some nations, like Brazil, are rich in natural energy resources; others, like Japan, have next to none. Some are in geopolitically secure positions like the United Kingdom; others, like China, are less secure. For those reasons, each nation will experience the break point, rallying cry, and rebalancing in different ways. Nevertheless, the rebalancing challenge we now face is more complicated than it was back in the 1970s. Oil has become more difficult and expensive to find, develop, and bring to market. Geopolitical issues still antagonize. Environmental and social pressures are more acute. And as we learned in Chapter 5, large-scale, "magic bullet" cures are not handily available to diversify our energy mix and ease the burden off our oil dependency, nor will this period of global turmoil we're entering end quickly.

Evolutionary Phases of Break Point and Rebalancing

Some experts believe that energy break points are resolved when economies slow down. True, economic downturns do allow energy demand to take a breather and provide suppliers with the time needed to catch up. But economic cyclicality is different from fundamental rebalancing. Normal cyclical forces are generally not strong enough to resolve the severe imbalance associated with a disadvantaged fuel in the energy mix. External measures are required to initiate real evolutionary change.

There are four evolutionary phases that a society experiences during pressure buildup and rebalancing:

1. Complaining and paying up;
2. Conserving and being more efficient;
3. Adopting alternative energy sources; and
4. Making societal, business, and lifestyle changes.

Though I categorize these dynamics as phases, bear in mind that they occur only in broad chronological sequence, and that there is plenty of overlap. For example, meaningful conservation and efficiency most often requires business and lifestyle change. And, of course, complaining and paying up is pervasive.

Complaining and Paying Up

Everyone understands the complain-and-pay-up reaction at an individual level. As gas prices rise, we groan when we fill up at the pump, keep an eye on prices daily, and listen for answers when pundits talk about the problem. We blame politicians, the foreigners who control oil supply, the big oil companies, or others in our society. While we find this all irritating, the pinch is not yet great enough to bring us to make any fundamental changes to our lifestyle. After all, we're not sure how long the high prices will last, and at the end of the day we still need to drive our cars to work, light our offices, and heat our homes.

Corporate leaders do plenty of complaining, too, especially when their revenue projections get hit by higher than expected operating costs and their stock prices begin to fall. In the first quarter of 2005, Delta Airlines complained publicly that: "Historic high aircraft fuel prices are having a significant adverse effect on our financial performance." Poor financial performance is only tolerated for so long. Shareholders usually only forgive a CEO for a quarter or two before they demand accountability for a company's declining competitiveness.

The churn of complaint creates much political sound and fury. Helpless politicians blame their own favorite targets, while trying to appease an angry public that is demanding renewed access to cheap, reliable flows of energy. Corporate lobbyists get into the mix, pushing politicians for subsidies or other legislative actions that reduce energy costs. Yet any lawmaker worth his or her salt will tell you that when energy prices are rising, it's perilous to pretend that subsidies aren't putting money in one pocket by taking from another. In many cases, it pays politicians to take a wait-and-see position since in the complain-and-pay-up stage there are no easy or quick solutions that will have any positive impact, nor is there any political reward for taking the right long-term steps. After all, the most effective rebalancing solutions extend well into future elections, and any politician of vision can recognize that the accolades for helping to accelerate the change will likely accrue to someone else.

Conserving and Being More Efficient

As pressure buildup grows toward the break point, a nation typically turns to conservation measures when paying up and complaining starts to have long-term impact on disposable income and corporate profits. In particular, lower-income people and price-sensitive businesses are forced to conserve. In the United States and Canada, people of all income levels have grown quite accustomed to climate-controlled lifestyles, in which every room of the house or workplace is always kept at the perfect temperature. In other countries of the world, including many that are our technological and economic equal, simple conservation methods, like leaving rooms that are not in use unheated are standard behavior. Driving less, driving slower, car pooling, and taking public transportation are all conservation measures that may be inconvenient but can make a big difference on a nation's energy intensity.

Even so, such measures aren't enough to permanently reverse the pressure buildup. In that aspect, conserving energy is a bit like dieting. Once the weight is shed, most people drift back to old bad habits quite naturally. To instill permanency, some conservation measures can be legislated. For example, during the energy crises in the 1970s, the U.S. speed limit was cut to 55 miles per hour. This was (and still is) an obvious way to conserve fuel because the gasoline required to go faster increases exponentially with speed. High occupancy vehicle (HOV) lanes are another way to conserve by getting people to trade off the convenience of sharing a ride with a faster and more fuel efficient commute. Many countries, including parts of China, have gone so far as to institute odd-even license plate rules, limiting driving to any individual vehicle to every other day. Smog and congestion were the principal catalysts of odd-even plate schemes; of course, reducing emissions and energy conservation go hand-in-hand.

The heavy-handed way to achieve permanent conservation is through taxation. Many countries outside North America—Japan and the United Kingdom to name two—slapped huge taxes on gasoline at the pumps to force conservation after the last break point. Whatever your feelings about taxes, the gasoline tax approach in those countries, and others, has certainly worked. Figure 6.1 shows the energy mix for the United Kingdom. from 1960 to present. Note oil consumption in the United

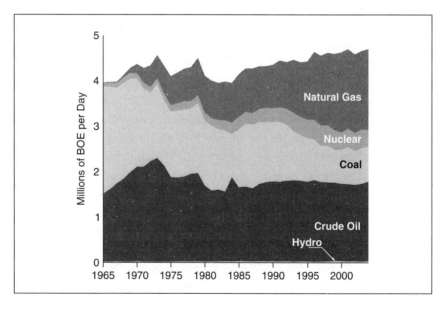

Figure 6.1   Evolution of the U.K. Energy Mix, 1965-2004: All Major Primary Energy Sources Converted to BOE (*Source: Adapted from BP Statistical Review 2005*)

Kingdom: It is actually lower today than it was in 1973. Fuel taxes did their job in helping to disadvantage gasoline permanently so that society—from industry to individuals—conserved and sought out alternatives. Perhaps the harshness of the early 1970's energy crisis helped solidify conservation resolve in British citizens and politicians. While the United States went through considerable pain at the gas pump, the United Kingdom had its energy crisis further exacerbated by a British coal miners' strike. Offices on Fleet Street even had to resort to kerosene lamps at one point, as though a century of energy progress had proved to be a dream. Later, the discovery of hydrocarbons in the North Sea further changed the United Kingdom's relationship with oil. The country that had once been foremost in striving to secure supply in the Middle East and around the world suddenly became a net exporter. Nevertheless, a curious thing happened as a result. With ample supply, and no longer any internal incentive to be efficient users of oil, the British government maintained its policy of high taxes, and society continued to conserve. As such the United Kingdom, along with other European countries and Japan that have largely decoupled their economic growth from oil demand, have the comfort of an economy that is less sensitive to changes in oil price than other nations like the United States, where the suggestion of higher gasoline taxes is political suicide.

To effect permanent change without relying on fuel taxes, a nation's focus has to be on energy efficiency. Efficiency can be measured up and down the energy supply chain. It can be measured specifically by looking at particular points, like the engine in your car; or it can be measured from end to end, as when we examine the efficiency of the chain from the oil well to your car. Again, only 17 percent of the energy in a barrel of oil typically makes it to the rubber on the road. If efficiency of the whole gasoline-powered transportation system was improved to 22 percent, a mere five-point gain, it would cut fuel consumption by 20 percent. In the United States that amounts to just over eight million gallons per day.

Whether it's the world's systems of road transportation, refrigeration, or lighting, getting a few percentage points more work out of a system can have dramatic effects on reducing demand growth. And the more inefficient a system is, the more leverage there is to saving fuel from small efficiency gains.

The ease with which efficiency can be improved at a national level depends on the type and placement of energy conversion hardware. For example, modernizing a big, clunky, old diesel power generator that serves 500 people improves the efficiency to 500 people's homes in one quick stroke. That's because the energy conversion hardware is centralized. Now instead of a centralized power system, imagine that those 500 people each had their own small and inefficient generators, in other words, a distributed system. Coaxing each person to buy a new, more efficient generator becomes more challenging and takes much longer to accomplish. This, in principal, is why improving the fuel economy of a nation's automobiles is so challenging; after all, cars and SUVs are a distributed form of transportation. Getting people to change vehicles is difficult, even with incentives, especially if they are on a tight budget. To accomplish meaningful improvements, everyone has to buy a new vehicle in short order. That's 230 million registered light vehicles in the United States, where a typical year's sales are only seventeen million vehicles.

Before the 1973 oil embargo, energy was perceived to be plentiful and endless, so there was no real incentive to conserve or be efficient. One solid outcome of the 1980s rebalancing was that the world became much less wasteful. Many of the changes became permanent as the old, installed base of hardware, vehicles, and appliances across the supply chains were rotated out in favor of higher-efficiency devices. In hindsight a lot of what was done was relatively easy. Getting rid of the big, heavy metal pieces off of Detroit's cars was not too difficult, and it meant less work needed to be done by the engine to drive the same distance. An appliance like a refrigerator used to be notoriously inefficient. To make people directly aware of the efficiency issue, the U.S. Congress made appliance labeling mandatory as part of the Energy Policy and Conservation Act of 1975. Public education helped us make smarter choices and gave appliance manufacturers a new, competitive dimension. A similar-sized refrigerator today uses 70 percent less electricity than its peer in 1974. Everybody came out a winner.

There is, however, a downside to this view that minor improvements in technology can save the day. Unlike consumer electronics such as personal computers, innovations in the world's energy supply chains occur less frequently and less dramatically. The world of energy conversion has its limits for a simple reason: The laws of physics dictate limits on efficiency for all energy conversion devices such as engines, turbines, and fuel cells. For example, an internal combustion engine like the one in your car or truck operates under the physical laws of what's called a thermodynamic cycle. It's the theory that relates to how much useful work can be extracted out of a combustion device like an engine, and how much of the input is lost to heat. A typical gasoline engine runs under a specific thermodynamic cycle called the "Otto Cycle," named after automobile pioneer Nikolaus Otto. Other thermodynamic cycles include the Diesel cycle, named after its inventor, Rudolph Diesel. Without diving into heavy theory, each thermodynamic cycle has a maximum theoretical efficiency. Mathematics formulated by French thermodynamics pioneer Sadi Carnot in the early nineteenth century allows any enterprising engineer to calculate those maximum efficiencies. Practically, the numbers are pretty grim: an Otto cycle gasoline engine can't convert much more than 25 percent of the energy in a gallon of gasoline into useful mechanical work; a Diesel cycle engine can not do much better than 40 percent. Today's high-tech engines are already pretty close to those limits. There is no Manhattan Project that can change the laws of physics.

If the internal combustion engine that is in your vehicle is close to its practical efficiency limit, then how can we possibly improve overall vehicle fuel economy? The answer to that comes from a question which I will pose as follows: "What do you define as useful work when it comes to a vehicle?" We know the input is a gallon of gasoline, but what is the desired output? The exercise is not merely spinning an engine and turning gears. The usefulness of a car is getting you and maybe your family, friends, and some cargo from point A to point B. There is no useful work being done when your vehicle is idling at a red light. Vehicles, like hybrids, that turn off at red lights obtain a large part of their superior fuel economy by preserving fuel when no useful work is being done. Hybrids are also good at recycling energy from braking back into motion. The energy in braking is captured electrically and stored in a battery. When you're ready to start up again, the battery delivers the recycled energy back to the wheels through an electric motor. In contrast, is hauling around 4,000 pounds of metal and plastic in a big SUV useful work? Is all this "extra baggage" a necessary part of the commute, when the really useful work is to get the average 175-pound individual from home to office and back?

Driving lighter vehicles is a very effective and low-tech way of improving fuel economy. Continuing our discussion about vehicles in Chapter 2, let's take a closer look at the issue of road fuel consumption in the United States.

Slowing down the rate at which a nation consumes road fuel is neither easy nor quick, and it's especially difficult if it's going to be accomplished in a manner that's lifestyle neutral. By lifestyle neutral, I mean carrying out a social change without altering the way people buy or drive vehicles. In North America, the issue of an individual's right to buy the car or truck of his choice is about as contentious as gun control. And a lot more people own vehicles than guns.

The difficulty of this challenge is highlighted when you consider that since the Model T was introduced in 1908, there have only been a handful years where road fuel consumption has stayed level or gone down in the United States. Not surprisingly, those years were mainly during the 1970s oil price shocks.

Without forcing people to buy more fuel efficient cars, there are two ways of curbing road fuel consumption: getting people to drive less, or getting them to use less fuel while driving. It sounds simple enough, but remember the political challenge is to do it in a lifestyle neutral way. So getting people to drive less than the average 12,000 miles per year is pretty tough now that a substantial portion of the population commutes by car—usually alone—to and from sprawling suburbs. After all, it's not realistic to think that masses of people are going to move closer to their jobs just to save a few bucks a week on gasoline. Building more mass transit would help, but it would be prohibitively expensive, take a long time, and violate our principal assumption that the solution cannot alter peoples' lifestyle. Carpooling is not in the solution set either, because it violates the same basic assumption. Ditto, raising fuel taxes to curb demand.

So, in the absence of legislated change, Americans are confined to the other solution: trying to use less road fuel for the same distance traveled. There are four options that serve to improve the fuel economy of vehicles on the road:

1. Reduce the average weight of the vehicle;
2. Switch to a fuel type that gets better fuel economy or displaces the crude oil supply chain altogether;
3. Reduce the average highway driving speed; or
4. Improve engine and drive train technologies.

Let's take a closer look at each of these possibilities.

Weight Reduction

Not including big transport trucks there are 230 million registered vehicles in the United States, 92 million of which are light trucks (pickup trucks, SUVs, and vans) and 138 million of which are conventional automobiles. There is a strong linear relationship between vehicle weight and fuel consumption. In city driving, fuel economy improves by about an average 5.6 miles per gallon for every 1,000 pounds of weight reduction on 2005 model vehicles. Midsize cars are about 1,500 pounds lighter than trucks and SUVs, so gains in fuel economy that result from switching to smaller vehicles can easily exceed 50 percent, going from say, 16 miles per gallon to 24 miles per gallon.

Therein lies one of the great ironies of today: Technology has offered us "magic bullets" to mitigate supply chain pressure and rebalance our energy systems in the past, yet the world now—

and America in particular—doesn't need a "technology ticket" or a Manhattan Project to cut oil consumption by a large margin. Convincing people to buy smaller, lighter-weight vehicles is the easiest way of improving a nation's overall average fuel economy.

Of course the problem is that a step down to a smaller vehicle is not always lifestyle neutral. Going smaller means vehicle owners lose the flexibility to transport a maximum amount of people and cargo on demand, an option for which society is paying a huge premium. There is also a strong belief that larger vehicles imply greater safety in the event of collision, another big obstacle in trying to convince people to "lighten up."

Fuel Switching

Switching to a fuel other than gasoline holds substantial promise, especially if that fuel is diesel. At present, only about four percent of the SUVs, pickups, and vans consume diesel fuel, and less than half-a-percent of cars. Yet, diesel engines can yield up to 60 percent more fuel economy than their gasoline counterparts. So what are we waiting for? Unfortunately, diesel is still thought of in the United States as difficult-to-start engines that cause black smoke to belch out as they rattle away from a stop light. While this used to be the case, great advances have been made to clean up diesel's act. Today's turbo diesel cars perform similarly to their gasoline-powered brethren, so much so that if you blind test drive one, you would have difficulty picking out the diesel until you saw how much you saved at the pump.

Higher emissions of nitrous oxides, the key ingredient of smog, are problematic, so diesel vehicles generally don't pass the scrutiny of the environmental lobby or tight air quality legislation in some regions. In fact, diesel vehicles are not available for sale in several states, including New York, Vermont, Maine, and Massachusetts. Technologies that clean up the diesel exhaust are available, but the more the emissions are cleaned up, the less efficient the engine becomes. This is a great example of how society must make difficult choices going forward. Increasingly, we are being faced with a choice between clean energy or cheap energy.

Finally, a practical drawback for drivers is that diesel fuel is not as readily available as gasoline in North America. Not all gas stations sell diesel, so this presents an inconvenience factor when filling up. In addition, diesel engines are slightly more expensive, presenting another hurdle in a new car buyer's decision. But in comparison to other alternatives like car pooling or odd-even license days, getting people to switch to diesel vehicles—with the result of an immediate, 25 to 30 percent pickup in fuel economy— is close to being lifestyle neutral. The Bush administration has picked up on this and is encouraging automakers to produce a new generation of modern, clean-diesel cars and trucks.

Reduce Average Driving Speed

In 1974, during the energy crisis, the U.S. Congress cut the highway speed limit to 55 miles per hour, down from 75 miles per hour in most States. It depends on the vehicle, but an average car traveling at 55 miles per hour consumes 17 percent less fuel than one traveling 20 miles per hour faster.

In 1987 the limit went back up to 65 miles per hour. In 1995, the federal government relinquished its jurisdiction over speed limit controls altogether, leaving the decision up to the states once again. Today, highway and freeway speed limits vary from 55 to 75 miles per hour, but it's a challenge to find anyone that drives less than 70 today. There is no doubt that reducing speed reduces fuel consumption; however, it's far from a lifestyle neutral solution. The last thing suburban drivers want is to increase their now lengthier commuting time by slowing down their driving speed.

Technological Advances

That brings us to our fourth and final option: building a better mousetrap.

Engineering better engines and drive trains with improved fuel economy has been an ongoing endeavor, and it's getting increasingly difficult. The last really major advancement was fuel injection, which was rapidly adopted throughout the 1980s. Hybrid vehicles, like the Toyota Prius, are the next highly touted leap.

Hybrid vehicles are powered by a combination of a battery-powered electric motor and a gasoline engine. These vehicles don't need to be plugged into an AC socket for recharging. The battery is recharged during use either by the gasoline engine, which acts as a generator, or by capturing the energy of braking. Instead of losing all the energy that goes into stopping a car into heating up the disc brakes, much of it is recaptured as electricity and stored in the battery for later use.

The systems that manage the balance between hybrid electric drive and engine drive are highly complex. Hybrids typically shut down at red lights and restart imperceptibly when you put your foot down on the gas pedal so there is much less fuel used during stop-and-go traffic. It makes sense: you only put the energy in a gallon of gasoline to work when it's needed. When driving on a free-way, however, the benefits are substantially diminished because a hybrid reverts to acting more like a normal gasoline-powered car. In a reversal of typical fuel ratings, a hybrid's city fuel economy is substantially better than on the open road.

So, while it's true that some hybrid models boast fuel econ-omy that's two-to-three times better than a nonhybrid vehicle of equal weight, that's only in the city. Someone who buys a hybrid vehicle will experience improvement in fuel economy that is highly dependent on her driving habits and commuting pat-terns. There are no aggregate road statistics on hybrid vehicles yet, but as a vehicle group it's likely that they will yield a 25 to 30 percent improvement in fuel economy over standard vehicles of equivalent weight. Hybrids are marvels of engineering and their cost and performance will improve over time as more peo-ple make the choice to buy them. Will hybrids stop the coming break point in its tracks? Not by a long shot, but for the next decade or two they will make imminent sense, dramatically improving fuel economy in the city while allowing drivers to continue indulging in the most entrenched standard in the energy industry—gasoline.

So here are a few simple alternatives to improving fuel economy by at least 25 percent with minimal compromise to lifestyle. You can buy a new vehicle that is lighter by 750 pounds (340 kg), burns diesel instead of gasoline, or is a hybrid.

But will your new purchase have a serious impact on national fuel consumption? To answer that question, we must solve the following problem: "How many high-fuel-economy vehicles must be sold each year to offset yearly growth in gasoline consumption?"

It's not an easy question to answer, because there are so many variables to consider. Remember that new vehicles are brought into the fleet each year, and old ones are retired. The overall fleet is expanding, and the composition between light trucks and automobiles is changing. By "high-fuel-economy" vehicles I mean those that have consumption ratings that are at least 25 percent better than a standard vehicle of equivalent weight. For example, a Ford Escape SUV reportedly has fuel economy of 25 miles per gallon, whereas its new hybrid sibling reportedly gets 33 miles per gallon, a 32 percent improvement. Buying a diesel versus a gasoline version of a pickup truck also qualifies. Assuming that lifestyle cannot be compromised simplifies the problem a lot, because we can extrapolate current trends where appropriate and assume status quo on many of the variables.

So if starting next year, each of the eight million Americans who are in the market for a new light truck (pickup, SUV, van) buys one that is at least 25 percent more fuel efficient than the fleet average of 17.8 miles per gallon, then gasoline consumed by the light truck segment will remain level with this year at just over 60 billion gallons per year. Now consider cars. If, starting next year, each person who is in the market for a new automobile buys one that is at least 25 percent more fuel efficient than the fleet average of 22.2 miles per gallon, then gasoline consumed by the automobile segment will still continue to marginally *increase*, though at a much slower rate as compared to the current 1.8 percent.

The reason that the light truck segment's fuel growth slows down quicker than the car segment is because the base of light trucks is smaller and the "churn" of new light truck sales is faster than that of cars; that is, people who buy trucks generally seem to replace them sooner than those who buy cars. If people started replacing cars faster, then the car fleet's average fuel economy would improve at a fast enough pace to offset growth. One way of promoting this is to provide financial incentives to buy a new car. In June 2005, President Bush proposed that every American who purchases a hybrid vehicle receives a tax credit of up to $3,400. This will help increase the churn in car sales, however, *every single new car sold from today forward in the United States needs to be a hybrid if gasoline consumption is to stop growing.*

Not very realistic, but let's go with it for a moment. Even if every new car and truck sale in America going forward were a hybrid, it would be at least a decade before our national gasoline consumption would reverse its current growth trend. This is not to discourage you from going out and getting your new hybrid, diesel, or lighter vehicle tomorrow. Just realize that it's unrealistic for every new vehicle buyer to do so in the absence of much, much higher gasoline prices or legislation. During the last break point, it took a combination of both to do the job, and even then gasoline consumption continued to increase steadily, albeit at a slower pace.

The hurdle is very high. Gasoline consumption by the growing fleet of U.S. automobiles and light trucks is extremely difficult to offset over the next decade without big changes to driver habits and lifestyle. Hybrids and clean-diesels will contribute to the solution, but are far from a near-term panacea. In the United States, there are no quick or easy ways to curb the growing demand for road fuels. In China, in contrast, the solutions potentially have more scope. Although China's oil consumption ranks second in the world, as a society it is still far less shackled to the petroleum standard than North America or Europe. In other words, it is easier going forward for individual consumers in China, most of whom will be first-time car-buyers, to make high-fuel-economy choices. Furthermore, it's not inconceivable that a more authoritarian country like China could mandate buying choices, raise fuel taxes, or outright restrict the use of fuel-inefficient vehicles. In short,

China, with only a small fraction of its population behind the wheel, still has the luxury to alter the course of how its driver choices and habits form.

Adopting Alternative Energy Supply Chains

When complaining, conserving, and being more efficient aren't enough of a remedy, adopting alternative energy supply chains becomes necessary.

Historically, alternative energy supply chains have fortuitously come in the form of "magic bullets," radically new ways of harnessing power from fuels, often from brand new energy sources. To name some obvious examples, coal and steam engines, rock oil and kerosene lamps, uranium and nuclear power plants. We know there are no such analogous magic bullets forthcoming anytime soon, and certainly not before the next break point, so let's discuss the push that's taking place to unpack some old magic bullets and try them again.

In the United States, big initiatives are brewing from within the Bush administration to rejuvenate the nuclear power industry by building new power plants. It's going to be a difficult sell, as nuclear power plants have been reviled by the U.S. public since the Three Mile Island disaster in 1979. The Ukrainian experience at Chernobyl in 1986 only reinforced deep American anxieties about radiation and nuclear waste. Nevertheless, President Bush is trying to convince a nuclear-leery public that, "It's time for America to start building [nuclear power plants] again."

While President Bush may have more than one reason for jump-starting nuclear power in the United States, reducing oil demand growth is not one of them. Nuclear power plants generate electricity. And you can't put uranium fuel rods in your gas tank.

Today, oil comprises about three percent of the power-generating fuel mix in the United States. The nuclear trump card to displace oil in the United States was played in the 1970s and there's very little left to squeeze out. In truth, nuclear power will be necessary in the United States to grow electricity generating capacity—

another looming energy issue—but no one should think that U.S. gasoline prices will go down if more nuclear power plants are built.

The same logic applies to the renewed emphasis on coal use. Emissions from burning coal have been cleaned up substantially in the past couple of decades. The upside of coal in the United States is that it is abundant and can power the nation for a couple of hundred years, so there are no foreign dependency issues. However, like nuclear power, coal is currently used to generate electricity in the United States. And yet, that's not to say it can't become an oil substitute. Germany, in trying to find alternatives to rock oil for making road fuels in between the two world wars, built up infrastructure to make synthetic liquid petroleum products like gasoline from coal. The processes pioneered in the early twentieth century by German scientists Frederick Berguis, Franz Fischer, and Hans Tropsch are no secret today, but methods of refining petroleum from coal have historically been disadvantaged, in other words too expensive, relative to cheap sources of crude oil. Of course, when your back is against the wall, and security of supply is a key issue, such extraordinary efforts make sense. Germany experienced a break point during the First World War, and fear of future scarcity in the interwar period motivated them to rebalance with alternate sources of primary energy from which gasoline and diesel could be made.

Though cheap oil in the end trumped coal as a source of gasoline, the German-developed petroleum-from-coal technology later found its way to another country with its back against the wall. South Africa, because of its policy of apartheid, found itself isolated from the world after international trade sanctions severely reduced oil supplies to the nation, creating a break point. In 1986, South Africa's then President P. W. Botha contemplated the damage: "Between 1973 and 1984 the Republic of South Africa had to pay R22 billion more than it would have normally spent. There were times when it was reported to me that we had enough oil for only a week. Just think what we could have done if we had that R22 billion today . . . what could have been done in other areas? But we had to spend it because we couldn't bring our motor cars and our diesel locomotives to a standstill as our economic life would have collapsed. We paid a price, which we are still suffering from today."[1]

To help the country rebalance, Sasol, a South African chemicals and fuels company, innovated on the Fischer-Tropsch process and built plants to produce petroleum products from the country's abundant coal reserves, and later their natural gas supplies too. Today, Sasol remains a world leader in synthetic petroleum technologies, and may hold the key—or one of the keys—to helping us rebalance our upcoming oil break point.

Petroleum products, like diesel, can also be made from natural gas, or indeed any other raw hydrocarbon, using technologies based on Fischer-Tropsch. The cost of making diesel from natural gas has historically been prohibitive too, but the economics are not quite as onerous as making petroleum products from coal. The technology is adaptable to a broader set of alternatives called "gas to liquids" or GTL for short. In fact, one of the largest energy projects in the world today is a GTL plant in Qatar. A small Arab state adjacent to Saudi Arabia, Qatar sits on top of one of the biggest natural gas fields on the planet, but it is far away from the big consumer markets and so is considered "stranded", because it's too costly to build pipelines to move it out to consumers. Consequently, Qatar, in conjunction with ExxonMobil, Shell, and others, is putting on a big push to convert its natural gas into petroleum liquids like diesel fuel, and then transport the products to markets around the world via tanker. The $7 billion project in Qatar is slated to start delivering 154,000 barrels per day of petroleum liquids, including 75,000 barrels of clean diesel per day, to Western markets by 2011. It's a nod toward rebalancing, but in the context of 9.0 million barrels per day of gasoline consumed in the United States alone, the massive Qatar GTL project will amount to not much more than a sip or two in our daily thirst for oil.

More promising is the ability of Qatar and other natural-gas-rich countries to turn natural gas into a cold liquid[2], called liquefied natural gas, or LNG. In this form it can be transported to ports around the world that are equipped to turn the liquid back into a gas. The potential is significant, because many countries are also expanding their use of natural gas in buses, vans, and other vehicles. Natural gas is not as robust as a fuel like gasoline or diesel, because of it gaseous state. However, it burns much more cleanly, giving it major appeal in smoggy urban centers, especially in Asia.

LNG has been a big part of rebalancing a whole host of nation's energy mixes since the 1973 oil crisis. First to really recognize its potential was Japan. Figure 6.2 shows Japan's historical energy mix.

Up to 1973, Japan was rapidly growing its top-line energy needs using oil primarily and coal to a lesser extent. The introduction of nuclear power and natural gas starting in the early 1970s can be seen clearly. Because Japan has no domestic natural gas, all the gas was brought in by newly constructed LNG tankers—a powerful demonstration of how building infrastructure for new

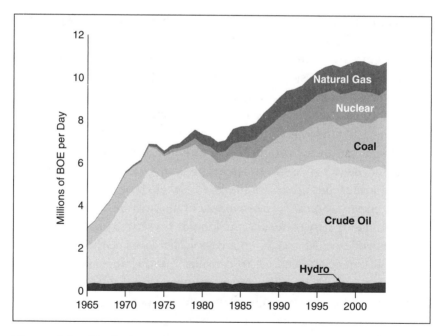

Figure 6.2   Evolution of the Japanese Energy Mix, 1965-2004: All Major Primary Energy Sources Converted to BOE (*Source: Adapted from BP Statistical Review 2005*)

supply chains facilitates rebalancing. Japan had a second wave of economic growth between the mid-1980s and mid-1990s. Note how the bulk of that economic activity was incrementally fueled by the growth of nuclear and natural gas in the nation's energy mix. By the way, if you go back to Figure 6.1, the United Kingdom's historical energy mix, you can see how natural gas also substituted for coal and oil. As noted earlier, neither the United Kingdom nor Japan consumes more oil today than they did back in 1973. And though both nations have had their economic ups and downs over the past 30 years, long-term energy policy has largely dissociated oil consumption from economic growth.

In finding different ways to refine petroleum products, there is a smorgasbord of alternatives between natural gas and coal, the two ends of the hydrocarbon spectrum. Natural gas in its purest form is called methane, the simplest, cleanest-burning hydrocarbon. Methane contains one carbon atom, surrounded by four hydrogen atoms. Coal, on the other hand, is a chemically complex solid mix of dead plant matter containing carbon, varying amounts of hydrogen, and countless other organic compounds. That's why coal-burning emissions are so unsightly and unhealthy.

Rock oils are nestled in between natural gas and coal in carbon complexity, spanning a range from light sweet crude to tarry bitumen. It's the job of refineries to distill raw hydrocarbons like rock oil into purer products like gasoline, jet fuel, and diesel. Chemically, it's possible to refine any one type of hydrocarbon into another, but processes like the Fischer-Tropsch are a much more energy-intense, expensive way of making diesel, than using a fuel like light sweet crude. As long as there was plentiful light sweet crude available at $20 per barrel, there was neither the need nor the financial incentive to consider refining fuels from costlier processes. But as the price of light sweet crude has risen, the dynamics have started to change. At $50 per barrel for oil, the alternatives of stranded natural gas and coal start to make economic sense.

And that brings us to an important point. What exactly is an alternative fuel? We are accustomed to thinking that when gasoline and diesel fuel get too expensive we have to immediately make a leap to new age solutions like hydrogen. But there is more than one way to make a fuel like diesel. The spectrum of alternatives increases as the price of light sweet crude goes higher, opening up the way for competitive refining processes from a varied range of dirtier carbon compounds.

This possibility further complicates the debate surrounding Hubbert's Peak, discussed back in Chapter 4. What resources should be included under the area of the bell-shaped curve in Figure 4.16? Is that area the exclusive territory of conventional crude oil, or can we also include nonconventional grades of heavier oils too? Should we throw in all of the earth's resources—from coal to kitchen grease—that can be converted into petroleum products? Doing so would also mean including agricultural products like corn and grain, which can also be refined into hydrocarbon fuels.

People who debate Hubbert's Peak are focused exclusively on whether or not the world's ability to produce oil has peaked. But at the end of the supply chain, people don't consume oil; they consume petroleum products like gasoline, diesel, and heating oil. Depending on price, all these petroleum products can be made from a multitude of carbon-based primary sources spanning natural gas to animal fat, which should all be considered alternative fuels. While I believe Hubbert's Peak really does describe the situation of light sweet crude oil, the debate is moot because alternative supply chains can emerge to make the petroleum products that people really want. I'm not saying that will be easy or cheap to do, or that it can happen quickly. But it can be done.

Perhaps the best example of higher price facilitating the emergence of alternative supply chains is the Canadian oil sands. In the wilderness of Western Canada there is a new-age gold rush going on.

Fortune seekers from the world over are flocking to Fort McMurray, Alberta, where billions of dollars are pouring in to extract bitumen from gooey, heavy black sand appropriately called tar sands or oil sands. Bitumen is the thickest form of rock oil and the last complex carbon stop before coal. Most of today's

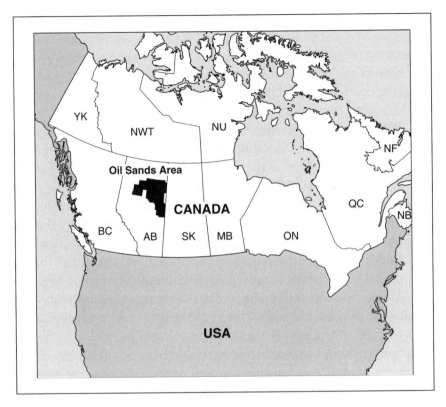

Figure 6.3   Map of Canadian Oil Sands Area: Showing the Peace River, Athabasca, and Cold Lake Areas

refineries can't handle bitumen, because nature's underground refinery hasn't cooked it enough over millions of years and turned it into the most coveted light sweet crudes. In the hinterlands of Western Canada, nature's course is being accelerated. After being separated from the tar sands, bitumen is "upgraded" into lighter synthetic oil blends that mimic WTI, and can then be fed into special refineries.

Like other oil resources the world over, the knowledge or exploitation of Canada's tar sands resource is nothing new. European explorers wrote of their experiences in seeing "bituminous fountains" in the late 1700s. Boring for bitumen began in the area in the 1910s. The first extraction plant was built on the banks of the Horse River, three miles from Fort McMurray, in the 1940s.

But it really wasn't until a company called Suncor established operations in 1967, and Syncrude Canada opened a plant in 1978, that noticeable volumes of synthetic oil started flowing out of the region.

Fort McMurray was put on the international stage in 2004, when oil economists at the U.S. Department of Energy officially recognized that Canada's tar sands contained over 200 billion barrels in oil reserves, the second largest accumulation of oil in the world, after Saudi Arabia. There is some debate about the overall size of Canada's oil sands reserves, because unlike conventional light oils, which are produced from oil wells, oil must be separated from the oil sands in a costly, energy-intense process. Nevertheless, John Cadman's 80-year-old prophecy about having to migrate to more expensive, secondary sources of oil has arrived and the resource is being aggressively developed.

The gold rush of investment in the oil sands, about $70 billion dollars over the next 10 years, is expected to nearly triple Canadian oil production from the Fort McMurray region to 3.0 million barrels per day by 2015. When oil was 20 dollars per barrel, it took vision by companies like Suncor to make investments in an area that was almost totally disadvantaged against cheap Middle Eastern oil. Everything changed when light sweet crude rose above 35 dollars per barrel on a sustained basis. Because of that price rise, oil companies now have incentive to accelerate the development of this new resource—an example of rebalancing the energy mix in action. Though it all began in the west with Kootenai Brown and John Lineham in the late 1800s, Canadian light oil didn't really start flowing in meaningful quantities until the 1940s. By 1975, after decades of pumping, Canada's conventionally drilled oilfields—those light oils that flow easily out of a drilled well bore—were declining, not unlike many other aged major reservoirs in the world. In other words, light sweet crude oil has peaked.

But total top-line Canadian oil production is rising due to the substitution of heavy "nonconventional" bitumen and synthetic oils (Figure 6.4) from the oil sands.

Venezuela too has huge oil sands reserves. However, the political climate there puts a premium on foreign investment, meaning

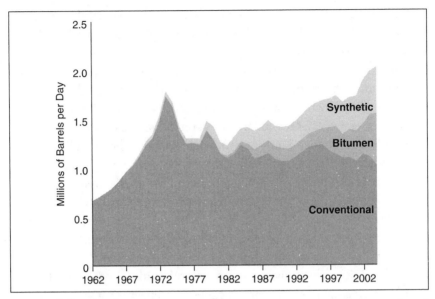

Figure 6.4   Total Canadian Oil Production, 1962-2004: Conventional and Nonconventional Sources (*Source: Adapted from* CAPP Statistical Handbook)

that Venezuela's barrels are effectively more expensive than those in Canada. The United States has huge reserves of heavier oil trapped underground in formations called oil shales. Abundant in places like Colorado and Utah, the issue again has been economics. An upgraded barrel of oil from the oil shales could not compete against a $20 barrel of abundant light sweet crude. But as oil prices have risen substantially, oil shales, too, are garnering a lot more interest lately. Security of domestic supply adds to the interest. The recurring issue, however, is time to market. Big, big dollars have to be invested, infrastructure has to be built, and such projects have to pass environmental muster and resistance from the not-in-my-backyard lobby. Canada's analogous oil sands investments started in earnest 40 years ago and only now are producing meaningful quantities. While it's likely that U.S. oil shales will be progressively developed over the decades to come, the chance that it will make any difference to America's national security or high-priced gasoline in the next 10 years is nil.

More likely is the expanded use of agricultural feedstocks to produce petroleum products. Synthetic diesel fuel, termed biodiesel, can be made from soybeans. And ethanol, which is a pretty close gasoline substitute, can be made from corn and grain. The beauty of these processes is that the feedstocks are renewable and the end products are free of sulfur, making them, in other words, "sweet." Also, agricultural feedstocks are grown domestically, so it relieves foreign dependency. President Bush proudly acknowledged this alternative fuel source when he said, "we're pretty good about growing corn here in America." The downside is that there is a long way to go before American corn growers can make a big dent in the 9.0 million barrels consumed by American drivers. On top of that, a large amount of arable land has to be devoted to fuel production. Are people prepared to pay more for fuel as well as bread, corn, and beans? Putting the scale into context, in 2004, agriculturally based fuels accounted for less than two percent of road fuel consumption. At current high gasoline prices, and with government subsidies under the new U.S. energy plan, agriculturally based biofuel supply chains will gather momentum, but the challenge of offsetting oil consumption is still daunting, because it's still really tough to beat the compelling utility of gasoline.

A much-touted alternative fuel is E85, which is 85 percent ethanol and 15 percent gasoline. Advantages of E85 relative to gasoline include a 25 percent reduction in exhaust pollutants, a 35 percent reduction in greenhouse gases, a 5 percent increase in horsepower, and only minor modifications to vehicle engines. E85 sounds like a dream fuel and yet, nothing comes for free. The principal drawback of a gallon of E85 is that it has only 72 percent of the energy content of gasoline. That means you need 1.4 gallons of E85 to travel the same distance as one gallon of gasoline. The expected doubling of ethanol production to 7.5 billion gallons by 2012 is impressive, but once again perspective is required. Comparing barrels to barrels, 7.5 billion gallons per year translates into just under 500,000 barrels per day, or about 5 percent of today's U.S. gasoline consumption. Yet because a gallon of gasoline contains 1.4 times as much energy as a gallon of ethanol, the potential displacement is not 500,000 barrels of gasoline, but closer to 350,000, or 3.5 percent, instead of 5.0 percent.

So, in terms of straight fuel economy, ethanol is a regression, though an advance in gasoline substitution. That is, assuming the size of the gas tank doesn't change, a driver that fills up with gasoline once every 10 days will find himself going to the pumps once every 7 on E85, a minor, but noteworthy lifestyle compromise. The bigger hitch is that there are only 188 E85 fueling stations in the United States, about half of which are in Minnesota. That's compared to 165,000 gasoline stations across the nation. Vehicle manufacturers are therefore building flexible fuel vehicles (FFVs) that can burn either gasoline or E85. But that flexibility is accompanied with increased purchase cost. E85 is a step in the right direction, and one of many rebalancing forces that will come into play next decade. But people need to realize that agricultural feedstocks do not have sufficient scale to offset near-term demand growth—let alone reversing gasoline demand—to avoid the next break point.

In general, renewable sources of energy have huge appeal, because unlike rock oil, coal, uranium, and natural gas, they are a nondepleting resource. Many European countries like Germany, Denmark, and the U.K. have installed large bases of wind power to help fulfill their electrical power needs. There is a strong push in places like Canada and the United States to increase wind power too, which makes good sense for generating electricity. But while generating power through wind, solar, and geothermal energy is laudable, for the same reason as nuclear power, none of it will curb America's oil demand.

The other big issue with renewables is scale. A good-sized wind turbine has a capacity rating of 2 Megawatts. A typical coal-fired power plant is around 400 Megawatts and a nuclear plant often exceeds 1,000 Megawatts. That means 500 windmills equal one nuclear power plant. Right? Well, not exactly. Nuclear power plants can operate over 90 percent of the time, and wind only blows about 30 percent, so you actually need 1,500 wind turbines to put out the same amount of electricity as one nuclear power plant. Moreover, those 1,500 turbines can't all be in the same place if they're collectively going to be running more than 30 percent of the time. And just because it's a renewable energy source doesn't mean that people want wind turbines in their backyards. Wind often blows best in beautiful mountain valleys and offshore. Nobody wants their pristine views ruined, yet everyone wants cheaper electricity to power their lives. Once again, the choice between the lesser of two evils emerges.

One nuclear plant can power 250,000 homes; one good-sized wind turbine can serve less than a couple of hundred. Ultimately, that's the biggest drawback with renewable energy. It's not a large-scale remedy at a time when large-scale solutions are necessary. Things like solar panels on rooftops make imminent sense, especially in sunny climates, but like wind power, solar cannot be scaled to make a real difference in a nation's oil consumption.

Pragmatism is what we need, because when you look at the amount of oil used in the energy mixes of industrialized and industrializing countries, you get a sense of the scale of the problem. Each type of alternative I described can make a contribution to rebalancing energy mixes around the world, unhooking us from our economic dependency on light sweet crude. But time is of the essence. If this current break point is going to be rebalanced within any reasonable timeframe, the world will need more than alternative sources of energy. We will need to make permanent societal changes.

Making Societal, Business, and Lifestyle Changes

Alternatives will slowly emerge, new supply chains will gather strength as old ones die out, people and corporations will adopt more efficient hardware, appliances, and vehicles, but these rebalancing trends are measured in decades. To speed things up, we need to accept lifestyle changes that involve consuming less energy. If we don't make those choices consciously, other choices will be imposed on us by economics, government, or both.

The price of oil is going to be very vulnerable to spikes over the next 5 to 10 years. High prices for gasoline, heating oil, and other energy commodities will help curb demand and impose some lifestyle change, until of course the prices ease a bit. But the world can't expect permanent relief from cyclicality; it needs permanent change.

Governments that have had a vision to curb demand have imposed lifestyle changes as tough medicine in the past. In 1978, Japan was the second largest importer of oil in the world, because it had no oil or natural gas resources. This made it extremely sensitive to the spike in oil prices in the 1970s. With resolve, the Japanese government put into action the report of its Advisory Committee on Energy, which aimed to reduce its 75 percent dependence on oil to 63 percent in the mid-1980s and to 50 percent by 1990. In fact, after having set these goals and benchmarks, Japan proceeded to meet them. Today, Japanese citizens drive cars, but they pay a high tax for gasoline and at toll booths and must undergo strict annual inspections that stabilize fuel economy levels. Much of Japan has a climate that is as cold in the winter and as hot in the summer as any place in the northeastern United States, however, central heating and central air conditioning are rare, even for the affluent upper middle class. Instead, rooms are cold in the winter, with occupants warmed by portable kerosene burners or low tables with electric heating elements. In summer, air conditioners are used sparingly to cool isolated rooms. While those lifestyle choices seem restrictive and even technologically backward to North Americans, they are culturally tolerated in Japan and have done much to moderate energy use.

Imposing lifestyle change by legislating on high consumer fuel taxes or outright limits on consumption is tough medicine. In North America especially, no one wants to hear that they have to cut back on life's taken-for-granted luxuries. People don't want to take a crowded bus, when they can enjoy the private comfort of their own cars. People don't want to heat or cool only part of their 3,000- or 4,000-square-foot houses either. In North America, the level of affluence is such that it's going to take much higher prices before many people change their ways. But to really offset demand we need to see lifestyle changes today at the level that was accomplished in Japan and many European countries over the past three decades.

The benefits of lifestyle changes are considerable, however, because they make a nation's economy more competitive. In a world where trade is becoming increasingly global, reducing energy dependency, especially dependence on high-cost fuels, is going to become even more important over time. Moreover, this timetable will not be measured in decades, but in years as Asian economies continue to industrialize at a blistering pace.

One of the first rules in a competitive business environment is that the low-cost producer that is not overly exposed to a single supplier is always the winner. The metaphor translates perfectly to the national level. Yet the public in North America is not conditioned to think that way. Our politicians continue to perpetuate the belief that cheap fuel, clean environment, secure supply, discreet infrastructure, and competitive economy all go hand in hand, just as it has since the days of Reid Sayers McBeth. At some point politicians will have to 'fess up with the public. Some day, we will hear the rally cry around a cause of national competitiveness, and we will know that the break point has arrived—and rebalancing is about to begin for real.

Not Something to Be Left to Markets and Businesspeople

Pressure is building toward a break point, the point at which the world's entrenched crude oil supply chains will become permanently disadvantaged relative to alternatives and nations will have no choice but to rebalance their oil dependency. But we don't have to wait for the break point to take action. Rising oil prices over the past four years have started the ball rolling. A strong ramp up in the development of the Canadian oil sands, U.S. moves to subsidize biofuels, China's ambitions to build more than three dozen nuclear reactors, and the aggressive buildup in worldwide LNG infrastructure are all indications that rebalancing is already underway, albeit not fast enough to avert more oil price volatility.

The phase of rebalancing that a country is in depends on that country's affluence and the composition of its incumbent energy mix. In North America we are still solidly in the complain-and-pay-up phase. People in lower-income brackets are being hit by high gasoline, natural gas, and heating oil prices, especially in the winter. Accordingly, there are some people in the lower social strata that are well into the conservation phase. At higher-income levels, rising fuel prices are considered a nuisance if they are considered at all. Many executives and investors that I've talked to have not even begun to consider the idea of driving less or compromising size and perceived safety by buying a smaller vehicle. It's unlikely such thinking will catch on even if the price of gasoline in the United States jumps to $4.00 per gallon, the price you'd see if oil hits a $120 a barrel. In the absence of government policy, dependency on foreign oil to fuel vehicles will keep creeping up in the United States.

Four dollars a gallon may not get a well-to-do SUV driver to change habits, but it would have definite consequences for the vast middle class. Jobs would be lost to greater unemployment in industries that can no longer compete on cost. Corporations that are doubly burdened by being energy intense and energy inefficient—for example, an outmoded petrochemical plant, lumber mill, or steel foundry—will suffer consequences. This is where high energy prices have the greatest impact, because corporations produce goods and services that recycle dollars through the economy. A lumber mill produces wood products, which are then purchased by homebuilders that are then purchased by home buyers. Dollars are circulated and the economy grows. On the other hand, driving a personal vehicle is only productive to the extent that it delivers people to their jobs.

One of the best examples of large-scale rebalancing in the "Alternative Supply Chains" phase currently underway is the LNG boom in Asia, the Middle East, Europe, and Africa. As mentioned, many countries boosted the natural gas fraction in their energy mix to suppress oil demand and to fuel economic growth. Some like the United States did it with their domestic supplies of natural gas and piped in imports from Canada. Others like Japan had to do it with LNG tankers and infrastructure.

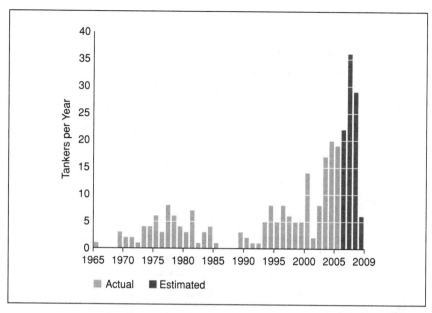

Figure 6.5   Construction of Liquefied Natural Gas Tankers: Actual to 2004, Estimated to 2008 (*Source: Adapted from Colton Company and ARC Financial*)

Figure 6.5 shows how many LNG tankers have been built in each year since 1965, about the time the technology to cool and transport LNG made its introduction. Note the rapid rise in tanker building activity between 1973 and 1985—the last break point and rebalancing act. Those tankers were largely destined for Japan. When the price of oil dropped, so too did LNG activity, because there was no need to build more expensive alternative supply chains. But look how activity has picked up recently, and look at the number of tankers that are scheduled to be built between now and the end of the decade. The tanker fleet is set to double in size, to about 350 tankers, by 2012. There is aggressive activity to build supply terminals in countries that have stranded natural gas, like Nigeria, Qatar, Trinidad, Australia, Iran, Indonesia, and others. Equally active is the construction of receiving terminals in China, India, South Korea, and Europe. In fact, about the only place where there isn't a lot of activity by comparison is the United

States, the country that needs it the most to facilitate further diversification away from oil in the energy mix. (Not to mention the fact that domestic natural gas supplies in North America have hit a break point in their own right, but that's another book.) The environmental, homeland security, and NIMBY lobby groups are all allied against LNG receiving terminals as well. It's not important why; it is important that other regions of the world do not have such inhibitions and are therefore going to be gaining the upper competitive hand as they diversify their energy mixes toward less volatile alternatives.

China is, in many respects, already at a break point; however, rebalancing is under way too. The government recognizes that their economy cannot continue growing at a blistering 10 percent per year using oil alone. Policy action is in place as the country moves to diversify its energy mix. And, as I hinted previously, there is a lot of potential to rebalance in China. The situation is clearly illustrated in the nation's energy mix, shown in Figure 6.6.

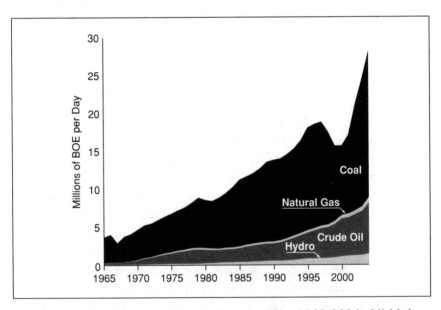

Figure 6.6 Evolution of China's Energy Mix, 1965-2004: All Major Primary Energy Sources Converted to BOE (*Source: Adapted from BP Statistical Review 2005*)

China's historical energy mix is not diverse. A little bit of hydro-electric power, a huge amount of coal, and a rapidly growing fraction of oil. Renewables are effectively nonexistent; nuclear power and natural gas are still relatively insignificant. Therein are the opportunities. China has put on a huge push to grow supply chains fueled by nuclear materials and natural gas. The government has a program in place to construct about 40 nuclear power plants by 2020. Western regions of China are virgin territory for finding large natural gas reserves, and LNG receiving terminals are slated to be constructed on the coast. Hydroelectric power will contribute more too when the third and final phase of the Three Gorges dam starts pushing electrons across the nation in earnest by 2009. If you compare China's energy mix today, it looks remarkably similar to Japan's pre-1973 (see Figure 6.2 for comparison). China has a lot of room to build up its energy mix such that its economy can grow without becoming excessively dependent on oil.

In addition, because China is already less shackled to the oil standard, it will find it easier to become more energy efficient (and less energy intense) in the future as well. As the pundits have pointed out, if every one of the 1.2 billion people of driving age in China buys a car and starts pumping the gas pedal in congested traffic, the nation's energy dependency would quickly enter the realm of the unsustainable. But the current pressure buildup is helping China accelerate its rebalancing agenda. If oil were cheap and plentiful today as it was in McBeth's time, China would be heading down the same addictive path as the rest of the world did. But resource constraints are encouraging rebalancing with sufficient urgency. While in North America we'll likely have to enter a "switch lifestyle" phase to tone down our voracious energy consumption, in China there is still a unique opportunity to avert gluttonous behavior. It doesn't appear to be happening yet, but at least the opportunity is recognizable. Neither region is in the "lifestyle" phase yet, but we're all well on our way.

The most compelling example of break point and rebalancing in modern times comes from South Korea, where a dramatic break point occurred in 1997. Figure 6.7 shows South Korea's energy mix from 1965 to today. South Korea's economy really started taking off in 1988, immediately after the Seoul Olympics. Trade was opened up and the country evolved rapidly into a manufacturing powerhouse. Supernormal economic growth subsequently averaged about 7.7 percent per year from 1988 to 1997. As you can see, the bulk of this growth was fueled with oil, the traditional economic rocket booster. On average, oil composed 60 percent of the country's energy mix. Indeed, South Korea's oil dependency factor was about a whopping 250 during this high growth period. Put another way, for every one percent change in the country's GDP, its oil consumption grew by 2.5 percent. Recall that in the heyday of economic growth in the 1960s, U.S. oil dependency was about 100 (one percent growth in oil consumption for every one percent growth in GDP). To push its rapid economic growth agenda, South Korea also began constructing nuclear reactors

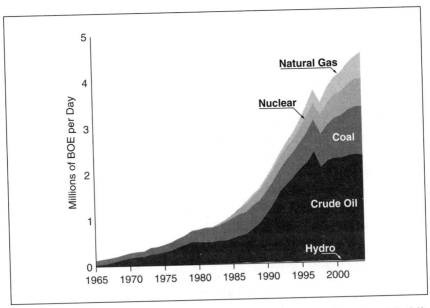

Figure 6.7   Evolution of South Korea's Energy Mix, 1965-2004: All Major Primary Energy Sources Converted to BOE (*Source: Adapted from BP Statistical Review 2005*)

and LNG terminals in late 1980s. Pressure gauges were looking good until things blew apart in 1997. Afflicted with a bad case of the "Asian flu," South Korea's currency rapidly devalued by 50 percent in 1997. Since oil markets are transacted in U.S. dollars, this meant that their imports of crude oil immediately cost twice as much. A sudden break point ensued.

Almost immediately, economic growth in South Korea contracted 6.9 percent. Demand for all energy commodities in the mix fell off, as you can see clearly in Figure 6.7.

It didn't take very long for South Korea to shake the flu and get back on a growth track, however. Since 1998 the country's economy has been growing by 6.1 percent, and its overall energy consumption is still growing aggressively as evidenced by the sharply increasing stack of primary fuels. But look at how the energy mix has rebalanced. Korea doesn't consume any more oil today than it did back in 1997. Its coal and nuclear use has expanded. Also notable is the large increase in natural gas consumption through the importation of more LNG. In fact, Korea is now the second largest importer of LNG in the world. By the time the nation recovered from its break point and rebalanced its energy mix, Korea's oil dependency factor dropped 90 percent to 20, less than half that of the United States today.

The rapidity with which South Korea emerged from its break point and rebalanced its energy mix is quite remarkable; indeed, it's the fastest that I've seen after studying the energy mixes of dozens of countries. But don't expect the same stunning changes in other nations. It takes a focused, national determination to accomplish what the South Koreans have done in so short a time period. Market deregulation and a conscious, government-directed effort to diversify the nation's energy mix was at the core of their directive. When examining Korea's overall energy consumption, the nation is still intense; however, as an ongoing extension of their rebalancing effort, the government is concentrating intently now on increasing overall energy efficiency. The South Korean example clearly shows how a rallying cry can catalyze rapid rebalancing in the face of a sudden break point.

The South Korean example also reaffirms a very important observation about countries that are in the throes of early, aggressive industrialization. Whether in Japan, the United States, Korea, or China, oil is the primary catalyst for rapid economic growth. The reasons are fairly nuanced. Oil is the most robust and flexible of all fuels. It's a liquid that is easy to transport and easy to store. Oil products are highly scalable, usable in tiny engines as well as huge engines. Oil products are also very robust in application: they can be used in leaf blowers, automobile engines, jet turbines, diesel power plants, blast furnaces, and home heating furnaces. No other primary fuel—natural gas, coal, uranium, or water (hydroelectric power)—boasts all these compelling attributes. Is it any wonder that oil helps catalyze growth? Is it any wonder that we become so dependent on this marvelous product? Is it any wonder that all nations who use this wonder product as a jet pack to propel their economies must ultimately learn to husband it and seek alternatives to rebalance and reduce dependence?

I think, by this point, it's become clear that reducing dependency and rebalancing a nation's energy mix is no easy task, though it can be done. If a rallying cry could originate from these pages, I would be the first to shout out. But blinded by our energy birthright in North America, and optimistic about the wonders of technology, too many of us are still convinced that help will come from some incredible magic bullet, the product of an unknown Edison working in a garage somewhere, ready to step forward when the world needs saving. In fact, wondrous technology will help save us. But not fast enough to help us make it through the next decade or two. In the meantime, alternatives to alternative energy will take the stage in the next great rebalancing act.

Thinking back to some of the break point and rebalancing examples that I have illustrated to this point, we can see that market forces alone are never enough to facilitate rapid rebalancing. In 1979 the planning Director of ENI, Italy's state-owned oil company, was quoted as saying, "Oil is a political commodity now. It is not something to be left to markets and businessmen.[3]" The statement is true for every historical oil break point, and I suggest it's true for any break point in the entire history of energy. How nations respond to the coming break point will necessarily go beyond the forces of mere markets and businessmen.

Notes

1 *Windhoek Advertiser*, April 25, 1986 (as taken from Richard Knight, March 2001 (richardknight.homestead.com/files/ oilembargo.)

2 Under normal atmospheric pressure, natural gas turns into a liquid at -260°F (-162°C).

3 *New York Times*, December 30, 1979 (as taken from Grayson, page 3).

CHAPTER 7

A GOLDEN AGE OF
ENERGY OPPORTUNITY

We've looked at what has happened to energy throughout history, and we've examined what's going on today. Now let's take a little trip into the future.

By the year 2017, the chaos and confusion of the previous decade had finally settled down. As with other break point and rebalancing periods in history, it was far easier in hindsight to see what had happened to our patterns of energy consumption and why. The clash of political and economic events that churned a froth of change in our life- and work-styles, had done much to create a brand new world. And yet, of course, there was much about this new world that was entirely familiar.

After our two sons had left home my wife and I relocated to the West Coast and moved into a townhouse on the edge of a small bay. I had an office on the upper floor with a panoramic view of the Pacific Ocean. Standing on the deck and looking down the coast some distance, I could see a large tanker pulling into a Liquefied Natural Gas (LNG) receiving terminal. I didn't have my binoculars handy, but if I had to guess, I'd say the ship was of Russian origin. The Russians had dominated natural gas deliveries to Europe for many years, and they had widely expanded the Eastern LNG export capabilities. No question the facility was a bit of an eyesore on the pristine landscape, but it was all part of the, "You can't have your cake and eat it too," reality that I had been lecturing about as early as 2004. Cheap, clean, secure, and discreet were four dimensions of energy that could no longer coexist. A decade ago, many communities across the continent had fought contentious battles to keep new refineries, terminals, pipelines, and power facilities out of their backyards, but the trade-off between inflating energy costs (rising at 10 percent per year locally) and a well-designed facility became easier to digest once wallets had been pinched and quality of life began to suffer.

Like many others, I believed that trade-off was worth it. I enjoyed my rural lifestyle. Between my ubiquitous Internet connection and the incredible advances in video-telephony, there was no practical reason for me to live near an urban center anymore. Telecommuting, the faddish buzzword of the 1990s, had finally gone mainstream as a result of further step changes in the Internet. The Web, still evolving, had become an "ether" of social and business interaction with an unprecedented feeling of face-to-face realism. From my desk with its ocean view, I had access to all of the data, information, and people I needed in order to do my work as an energy investment strategist and writer.

I wasn't the only one who had made a deliberate lifestyle choice in recent years. Some interesting demographic shifts emerged as an indirect consequence of the break point. Urban-minded people started moving closer to city centers, and distant suburban communities started to congeal into self-contained "3e-villages"—coined on the premises of ecology, energy efficiency, and electronically enabled—now commonly referred to as "Triple Es." Real estate analysts and demographers can tell you more about it, but in essence it was an evolutionary move against the long and costly commuting that had created so much incremental gasoline demand in the 1990s and early 2000s. By 2009 the number of miles clocked on the average American odometer peaked at 13,000; finally there was hope that gasoline consumption would at least level out in the United States.

The Triple E village concept had also catalyzed a whole new growth industry called community energy. There were hints of the trend germinating at the start of the century, but it really started gathering interest around 2011, well into the rebalancing period, piggybacking off the next wave of telecommunications technology. Triple E villages were actually an aggregation of dense, self-contained residential areas with services. From an energy perspective, the novelty was that the villages sought to be as energy efficient as possible and tailor-made their energy mix optimally for their locale. Effectively they were energy co-ops, and our local mix had evolved to include about 70 percent natural gas, and 30 percent biomass-generated methane for heating and stoves. Two large wind turbines down the coast combined with integrated solar panels on residential rooftops supplied about 20 percent of our local electricity, with 80 percent still having to be imported from the main grid. Gasoline remained the overwhelming transportation fuel, though 15 percent diesel was a notable change, compared to 2 percent 10 years prior.

I didn't miss my old commute, although I did miss the family SUV that we finally sold in 2008 when spiking gas prices, social pressures, and the rallying cry of the break point swayed us into trading in for a more responsible model. We should have done it earlier, but admittedly we were caught up in the broader public apathy. That apathy diminished quickly and social attitudes started changing toward the latter half of the decade. Believe it or not, a rally cry by politicians actually struck a chord with the public and many segments of the consumer goods industry looking for a megafad to piggyback new products on. It took a couple of years, and a few geopolitical standoffs like the deepening of tensions between Iraq and Iran to accelerate the agenda, but soon enough it became decidedly uncool to be an energy hog. Or, more to the point, it became cool to be energy efficient. Only the most unconscious or thick-skinned of us still rode around in anything resembling a Humvee.

Instead, I drove a lightweight, two-seater, diesel-powered, compact car most days, while my wife used her hybrid hatchback. As a car enthusiast and early adopter of new technologies, I was excited to soon be getting my new hydrogen-fuel-cell-powered convertible. It was a bit more expensive than I wanted to pay, and I was still a little concerned about the limited fueling options in our community, but after years of delay, and many overhyped promises, ToyotaGM had finally worked out the bugs and launched a line of vehicles that had the potential to hit the mark with the public. Of course, like other vehicles of today it will be networked wirelessly into the Internet's traffic flow management system, a major advance that has helped ease traffic congestion and reversed growth in fuel demand.

A new entrepreneurial venture, run by a fellow with infectious enthusiasm, was busy securing sites for hydrogen fueling stations all along the West Coast. We were fortunate enough to be close to one of the venture's early locations, which is why I chose to order my fuel cell vehicle. For the first time in a decade it really looked like the hydrogen business was going to gain some traction, and the charismatic CEO was already being touted as a visionary who had the shrewdness and take-no-prisoners attitude characteristic of Rockefeller, Getty, or Gates. To facilitate faster adoption, the CEO managed to convince hydrogen fuel makers, auto makers, and government safety authorities to collectively agree on fuel consistency and a standardized fueling nozzle that allowed for robotic fueling stations. It was Edison at his best. Finally we could say that hydrogen was officially in the transportation fuel mix, analogous to where electricity and lightbulbs were in 1895. It was a great start, but I know that it will still be a couple of decades before mass-market adoption kicks in.

All in all, it was a good life. After years of always feeling as though I were racing between meetings, conferences, and clients, things had finally settled down. I enjoyed the sense of clarity and calm I got from living in our small, self-contained community along the coast. Most days, I worked on my book about how recent, exciting advances in fusion technology were going to impact the coal and nuclear industries over the next half century. It had been 60 years since the world had seen a "magic bullet" in energy, and it finally looked like another one was on the foreseeable horizon.

But my book was not on the agenda for today. Instead, an editor at a national journal had asked me to summarize the last 12 years of energy history. Time to get at it. I took a last look out over the ocean, watching it undulate and flicker with the tide and the sunlight. It was easy, at such moments, to feel in touch with the planet. Our days of relying on fossil fuels had not gone by the wayside; in fact, we were more reliant than ever. But we had rebalanced our fuel mix through a variety of very large and very small economic and lifestyle shifts. The future looked bright. As I began to type my notes into the computer, some movement in the distance caught my eye. A pod of whales surfaced in the distance off the coast. I watched them with binoculars for the next 10 minutes, never tiring of seeing such magical creatures playing in the ocean. Then I got back to work, still amazed that we had once recklessly chased whales across the oceans for fuel.

How the Energy World Changed: 2005-2017

Break Point: 2007

Despite the buildup in pressure, nothing much changed in the world's attitudes towards energy up to 2005. Consumers in North America shrugged off ever-climbing gas prices and continued to buy SUVs and large cars with little regard to fuel economy, remembering the 1970s as a temporary blip in a long history of cheap fuel. Politicians blamed each other, foreign oil producers, and China, while the Federal Reserve Chairman's warnings about the consequences of energy price inflation on the economy only stirred up volatility in the stock markets without inspiring any coordinated or deliberate policy reaction. The world economy continued to grow at about four percent per year through 2006, and oil supplies were tightened severely as China, despite having to pay more and more to subsidize retail energy prices, continued to rely on oil as a "rocket booster" to fuel its supernormal economic growth. This left all importing nations in a precarious state, especially in the wake of Katrina and Rita's devastation, but it wasn't until the fall and winter of 2007/2008, that a confluence of events forced the public and our nations' leaders to finally realize how vulnerable we had become.

Hurricane Steve, another category 5 whopper temporarily disrupted Gulf of Mexico oil and natural gas infrastructure again, leaving heating fuel levels dangerously low in advance of what would be one of the coldest Novembers on record in the Northeast. An explosion at a Venezuelan tanker port caught fire in late November, leaving U.S. Gulf Coast refiners scrambling again for 750,000 barrels a day of crude. And in December, terrorists managed to blow up two supertankers in the Straits of Hormuz, leaving an environmental and geopolitical mess. Navies from all over the world flocked to this vital chokepoint, jockeying for position with no particular united command or strategy. You can imagine the rhetoric that played out at the emergency U.N. meeting. While tanker traffic was only disrupted for a couple of days, supertanker insurance rates went sky high and more naval escorts became part of the cost of doing business. One by one, the United States, India, China, Malaysia, Korea, Japan, and the EU announced formal intentions to accelerate strategic petroleum reserves. Analysts estimated that at least 1.2 million barrels per day were being diverted to these hoarding facilities.

In the United States, many rational people had been concerned about the pressure gauges rising for years. Hurricanes Katrina and Rita brought the issues into the focal point of people's wallets, but it took a series of natural and political calamities to really give the rally cry momentum. This wasn't surprising; as James R. Schlesinger, President Jimmy Carter's first energy secretary in 1977 aptly described our approach to energy as far back as 2005, "We have only two modes — complacency and panic.[1]" With the panic of 2007/2008, there was no longer anything left to debate. The U.S. president's national address calling for a "New Energy Dawn" and instituting sweeping regulations would later be used by historians to mark the turning point of meaningful change. And the United States was not alone as high-growth Asian countries adopted New Dawn type policies of their own to move into a serious rebalancing phase in the ongoing evolution of energy. The twenty-first century's first break point had finally been reached.

In fact, the break point had been looming for several years. The world economy stopped "firing on all cylinders" in 2007 and stagnated down to 2.5 percent growth per year for a few quarters. The demand for oil and the demand for all other energy commodities, like coal and natural gas, saw a temporary economy-driven slowdown in the tail end of the decade as a consequence of this stalling of the world's GDP growth. Although the price of oil pared back to less than 50 dollars for a few weeks, the world slowdown in economic growth didn't actually solve any of our oil-consumption issues. Instead, the vast majority of us recognized, as we had in the late 1970s, that rebalancing and conservation had to be accelerated if economic growth was to return.

Because of volatile oil prices and greater rebalancing efforts, oil demand fell more than demand for other primary fuels. Nevertheless, year-over-year growth for oil was still positive throughout the economic slowdown. Because supply was difficult to secure and strategic military issues came to the fore, pressure continued to linger and build. A growing hoarding mentality did not help matters. The great scramble for oil assets intensified as the world's "crown jewels," locations where the last of the "elephants" might be found, became largely spoken for by the end of 2007. The focus increasingly shifted to acquiring producing properties and consolidating independent oil companies. In particular, China, India, Russia, and several other Asian countries were the most aggressive in buying smaller western independents to secure supply and gain additional access to competitive knowledge.

The outcry over this "corporate pillaging" petered out as time wore on. Henri Berenger's much heralded quote that "He who owns the oil will rule his fellow men in an economic sense . . ." became an old-school doctrine. The new doctrine that emerged could perhaps be paraphrased as, "He who uses energy in the most efficient and productive manner will rule his fellow men in an economic sense . . ." Companies in the manufacturing industries, in particular, finally stopped complaining about high energy prices eating away at their profits. In part, this was due to the fact that no one was listening anymore, but it was also because those companies that adapted to the new energy realities assumed positions of unassailable competitive advantage. Just like the mantra in the 1990s had been for corporations to become lean and mean through use of information technology, so the mantra leading into 2010 was for companies to become more productive through smarter use of energy.

By 2010, the world was consuming about 90 million barrels of oil per day. Concentration of supply in the Middle East had become much more acute, causing intense price volatility. Though OPEC had *de facto* disbanded, the core cartel nations were investing at a steady pace in order to supply the world's call, but spare capacity remained tight. Billions were required every year to keep up with world demand, yet for many producers, like Venezuela, reinvesting oil profits back into the ground was not the first priority on the government's agenda. Non-OPEC countries became more important and more influential in terms of incremental supply, but the costs to get out the oil found at the "ends of the earth" escalated steadily. The world's oil barrels became heavier as light sweet crude became increasingly difficult to find. Oil sands production increased from 1.2 MMB/d to 2.2 MMB/d, principally from Canada. Iraq, after years of insurgency, was producing 3.5 MMB/d by 2010, up from 1.9 MMB/d in 2005, as investments in new infrastructure finally began to be effective. Though still mired in regional political issues and the on-again, off-again antagonism with Iran, Iraq was showing the potential to become a major regional player on the global scene with 7.0 MMB/d of oil production and growing by 2017. Still, over the years, the real power broker in the oil world had become Russia. By 2012 it boosted its production back up to 11 million barrels per day, or around pre-Berlin Wall levels. By 2015, it was producing more oil than Saudi Arabia, solidifying its position as the dominant producer of the twenty-first century.

Rebalancing: 2010-2017

Global energy intensity and the oil dependency factor started to decline after 2010. In other words, less oil was needed to transact each new dollar of economic growth. Looking back at the numbers, we can see that in the superheated years from 2002 to 2006, our oil dependency factor was close to 40, while in 2010 it had slowed back to 34 as market and policy forces began to accelerate rebalancing. It wasn't a major retreat yet, but the important thing was that it was starting to moderate. The contrast only highlights how unsustainable the economic fundamentals were in those remarkable years in the first decade of the millennium. As the world economy began to pick up again in the new decade, many of us were grateful that the break point of the latter half of the decade had clarified the need for new solutions and tangible approaches.

How exactly did we get there? In terms of world energy mix, a number of crucial changes had evolved. Environmental and climate change issues had become acute by 2010. The link between pollution, global warming, volatile weather patterns, and inefficient consumption of fossil fuels was finally gelling in the minds of the general public. Some argued that congestion and smog in Asian countries, particularly China, had been as much a factor in the break point as oil supply issues. The wave of environmental legislation in 2007 had crimped supply chains, keeping the pressure up. To industry insiders, this new pressure exacerbated supply problems, while more balanced assessments later pointed out that environmental lobbying actually facilitated a faster rebalancing. Among the many dynamics at play, more natural gas and renewable energy sources on the supply side, coupled with the adoption of smaller, more fuel efficient vehicles on the demand side, had the dual effect of rebalancing and satisfying environmental legislation. In addition, most mainstream environmentalists—especially those most closely connected with global warming issues—reconciled themselves to the renaissance in nuclear power, as 10 new U.S. plants were built or in the offing up to 2017. Outstripping the U.S. pace, China constructed 22 new nuclear power plants during the same period, helping to nudge oil out of its electrical power markets exactly like the United States, France, and the United Kingdom had done four decades earlier. India and other Asian nations were also aggressively building nuclear power plants. One big issue was a growing tension in the world's uranium supply chains, but because of renewed exploration efforts and a series of new mines scheduled to start producing in a couple of years, it looked like the pressure build could be held in check.

More disturbing to environmentalists was the rapid reemergence of coal to displace more oil out of the power generation markets in Asia. Growth in coal accelerated between 2004 and 2010, becoming the fastest-growing fuel until the end of the decade, when the increased use of natural gas helped ease the pressure. Yet coal consumption wasn't going down, and the increasing use of coal liquefaction technology, to produce synthetic petroleum products, was not welcome from a global warming perspective, because of the amount of carbon dioxide emitted during the process.

Coal prices rose steadily over the decade, and supply chain pressures could be felt in pockets around the world. The real beneficiaries of expanding coal use were the world's railroads and shipping lines that carried the commodity. Fortunately, clean coal technologies helped alleviate toxic emissions, but climate-altering greenhouse gases like carbon dioxide were still a big concern, particularly in parts of Asia, where lobbyists appeared to have little influence. China actually surprised many and became increasingly environmentally conscious even as it continued to expand its use of coal to help ease the burden of reliance on oil.

Globally, natural gas was the most aggressively growing fuel in the mix. Demand had tripled after the first oil break point and rebalancing of the 1970s, but by 2007 major LNG projects and new pipelines through Asia accelerated the growth of natural gas consumption. By the end of the decade, the world's LNG fleet and its LNG infrastructure had effectively doubled. Liberation of the world's "stranded" natural gas reserves was the substitution that had the most influence in rebalancing, much like nuclear power had helped us by substituting for oil in the 1970s. The United States was "late to the LNG party," but soaring North American natural gas prices in the wake of Katrina and Rita muted the not-in-my-backyard critics and accelerated permitting and construction of much-needed ship terminals. Natural gas was a global commodity on par with oil. The use of natural gas as a fuel in cars and trucks was climbing in many countries, especially in Asian countries where big-city smog was an issue. Students of energy history would recognize this as a classic growth and dependency phase for natural gas, but for now nobody was concerned about the addiction to this cleanest of the carbon-based fuels.

Ultimately, the planets had aligned by the end of the decade. The agendas of environmentalism and energy conservation had increasingly overlapped, and coupled with the volatile events of winter 2007/2008, everything had come to a focal point, triggering the break point.

All in all, by 2015, the world's energy mix had shifted to the point where we were less dependent on oil and more dependent on natural gas, coal, and uranium. Also, renewable energy sources were more than just a rounding error in world's energy mix. But there was lingering tension in the air, not only on the oil supply chain, but on the coal, natural gas, and uranium supplies, too. The absence of any "magic bullet," technology to relieve us of our fossil energy addiction kept the pressure up.

Fortunately, around 2015 developments in engineering and physics from the International Thermonuclear Experimental Reactor (ITER) built in France looked very promising. The $12 billion dollar project seeded in 2005 and pioneered by a six-partner consortium—Russia, Japan, South Korea, China, the European Union, and the United States—was starting to pay significant dividends. The leading team of scientists overcame some of the most perplexing problems of containing a miniature sun in a "magnetic bottle." Their achievements have been the stuff of Einstein, and this completely new energy supply chain originating from a feedstock of refined sea water (a resource that is truly abundant) is likely to be the most valuable magic bullet in our energy evolution. Such is the excitement that the first commercial reactor looks set to be built before 2030, although it will be at least a decade after that before we'll see notable changes in our tired and stressed energy mix. And it won't be a minute too soon, because Africa may be the next area of the world to rapidly industrialize, and it's not clear that yet another high-population region will be able to exploit fossil fuels, especially oil, as a rocket booster for economic growth.

In any event, today, in 2017, one thing in the world of energy has become clear: energy cycles are constantly evolving; the more things change, the more they stay the same.

Back to the Present: 2006 and Onward

My ocean view contracts, shrinking to the size of my screen saver. The office complex I work in comes back into focus, as my dream home on the coast retreats into fantasy. I look up from my work and around the room. Here I am, at my desk, keeping tabs on another crazy day in the energy markets. I have a dozen phone messages and e-mails from journalists and clients who want to know what's in store for the future.

Of course, part of me wishes that I had such an accurate crystal ball and could tell them what the world is going to look like in 2017, and what happens to us along the way. But another part of me does not want to guess what will happen, because I prefer to experience it in the moment. Many great developments will take shape in the next 10 to 20 years. New Rockefellers, new Edisons, new Einsteins will emerge. Nations will grapple with complex geopolitical and policy issues. New technologies will affect us in ways we can't yet imagine. No doubt, the progression of history will seem natural, if still remarkable, in hindsight, but the next 10 years will be filled with surprises around each corner we turn.

While I have no way of predicting the many events and developments that might easily alter our course, I do have a clear sense of the way our energy consumption patterns will evolve and how that will influence our lives on national, business, and individual scales. Because we know the mix of fuels we rely on today and the economics of getting those fuels to market, we can predict how those fuels will evolve. Because we know how entrenched and established our energy supply chains are today, we can rule out the introduction of any radical new technology that will change the nature of the game in the next 10 to 20 years.

So, what can we expect? We're in for a rough ride until the end of the decade, and probably for the next 10 years. Prices for oil and its associated petroleum products are likely to flare up annually, if not more often. Whenever prices start to ease due to short-term cyclical factors, everyone will heave a sigh of relief and think, "Thank God that's over; now we can carry on with our normal business." But the uncertainty and difficulty will not be over. Nearly every facet of society is dependent on petroleum products refined from light sweet crude. As a result, we will continue to need more and more. A thousand barrels a second is not a finish line where we get to rest after a long, hard race; it's just a milestone along the path to greater dependency.

The demand for petroleum products will continue to rise aggressively for the next several years. Moderation will not come until demand-side rebalancing starts to make a difference. The most notable factor will be China's ejection of oil as a rocket booster for economic growth. Any momentary kink in the growth of demand will be due to economic slowdown. While it is likely the world will not be firing on all cylinders economically for each of the next 10 years, we must be aware that natural rhythms in our economic cycles will not solve our problems. Price relief will be temporary and influenced by the seasons. Whether it's driving season, air conditioner season, hurricane season, or heating season, there will be constant tension between supply and demand and a bewildering sense of nonstop volatility. As a result, paranoia in the markets and a hoarding mentality in business, government, and society will prevail until such time that the world becomes satisfied that oil demand is no longer out of control.

Other factors will be at play along the way. Geopolitics will throw a match onto the fuel at a moment's notice. Pressure from environmental groups and NIMBY lobbies will combine with general public apathy about conservation to crimp the oil supply chains tighter, intensifying the flare-ups when they occur. At some point within the next few years, those flare-ups in price will start triggering break points in individual nations around the world, most likely starting with Asian countries. Then the great global rebalancing act will begin.

As I discussed in Chapter 5, rebalancing is already underway. But the really significant modifications will not begin in earnest until the latter part of this decade. At that time, the hundreds of billions of investment dollars being ploughed into new infrastructure will start to pay off. If geopolitical tensions greatly increase or the price spikes are high enough in the next few years, the agenda for rebalancing will be accelerated. Naturally, the faster prices and tensions rise, the faster we will be compelled to do something about it. Because the economics of energy are not going to improve for all of the reasons we've already discussed, government policies can make a big difference in how rapidly a nation evolves. Those nations that implement intelligent and visionary policies soon will be much better positioned to take advantage of the next period of robust economic growth, likely sometime in the middle of the next decade.

Where will the price of oil settle out after we reach the break point? The days of the $20 barrel are over. According to simple economics, the price of oil should settle out to the level of the cost of whatever it takes to bring the very last barrel of oil demanded to the market, the so-called marginal cost. That last barrel of light sweet crude is not going to come from a cheap, abundant field in Saudi Arabia; it's going to come from the ends of the earth, like spermaceti from the frigid Arctic waters in the waning days of the whale oil trade. Accordingly, the cost of oil is going to stay high, though we will be relieved when the volatility ends.

The examples from history throughout this book draw our attention to ongoing human efforts to find new supply chains of energy from which to turn our wheels, heat our homes, and light our cities. Through those stories, we can gain many lessons about what to expect over the next 20 years. Whenever I need to encapsulate those many lessons into one basic notion, I turn to the thoughts of Alfred Marshall, the famous nineteenth century English economist. In the fifth book of his *Principles of Economics*,

Marshall wrote: "At the beginning of his undertaking, and at every successive stage, the alert business man strives to modify his arrangements so as to obtain better results with a given expenditure, or equal results with a less expenditure. In other words, he ceaselessly applies the principle of substitution, with the purpose of increasing his profits; and in so doing, he seldom fails to increase the total efficiency of work, the total power over nature which man derives from organization and knowledge."

Marshall's principle makes obvious what we see around us today in the many goods and services we consume: If there is a better or cheaper way of doing something, we'll find it. No less important is the converse of Marshall's words: the human psyche is not inclined to pay more money for less product or utility. Newcomen, Boulton, Watt, Rivera, Sawyer, Edison, Otto, and countless other protagonists in the history of energy innovation have validated Marshall's principle. Inventors, businesspeople, financiers, and even promoters, have all helped to fill our energy needs in better and cheaper ways.

If I were to apply Marshall's principle to the economic challenges we face over the next 10 to 20 years, I would sum up the inference as follows: "Low-cost producers win." What does this mean? Notwithstanding the usual preconditions of quality, customer service, time to market, and so on, companies that operate their businesses at the lowest cost of their peer group take market share, are affluent in their profitability, and enjoy special status as darlings of industry. Nations that have aptitude in low-cost production enjoy similar rewards, but also accrue the benefits of having more economic power and political clout on the world's geopolitical stage. This does not mean that nations are able to passively wait for industries to lift them up; rather, nations in such an enviable position have usually had the vision to provide corporations with the policies and infrastructure that facilitate low-cost production. Meanwhile, individuals that focus on lowering their household energy costs will manage better financially and generally live a more abundant lifestyle. Collectively, those individuals contribute to the well-being of the entire society and the health and vigor of the culture.

For all of the reasons that we've discussed in this book, applying Marshall's principle to become a lower-cost consumer of energy is getting increasingly difficult. But Marshall also noted that, "If there is a better, cheaper way of doing things, humans will find it." Those nations, business, and individuals that can apply Marshall's principle more quickly and thoroughly will be at a significant competitive advantage going forward as energy prices continue to rise and grow in volatility. This signals a tremendous opportunity for those who focus on the new rules of the game to get ahead more quickly. Let's take a closer look at how that will work at the government, corporate, and individual levels.

Governments

Oil-importing governments who want to get ahead need to set proactive energy policies that slow down their nation's energy consumption. The solutions might be initially unpalatable for the public (particularly if that public is not aware of the issues), but visionary policy makers who take action today will help create a more globally competitive nation tomorrow. Educating the public and having them rally around the cause for the greater good is key to success. Policies that don't deal with near-term issues head-on will continue to leave the real power in the hands of oil exporters.

Of course, oil-exporting nations cannot be expected to alter their policies for the altruistic benefit of the rest of mankind. Each major oil exporter, whether a member of OPEC or not, has its own strategic intent in terms of how to exploit its oil reserves. Energy policies of these major exporting countries range from market-driven and yielding, as in Canada, to arbitrary and belligerent, as in Venezuela. Other exporters can be characterized as cagey and calculating (Russia), cleverly opportunistic (Libya), and measured and guarded (Saudi Arabia).

At the heart of it, we all need to recognize that in this global economy and relatively borderless world, there is no collective ambassadorship or world energy policy to save us from the planet's energy woes. Each producing nation has its own selfish agenda when it comes to the monetization of this increasingly valuable commodity, and each consuming nation has its own selfish agenda on how to obtain and exploit this commodity. The end game will be won by the most prudent and intelligent global consumer: "He who uses energy in the most efficient and productive manner will rule his fellow men in an economic sense. . . . "

The United States

Three out of four sectors of the U.S. economy—industrial, commercial, and residential—now have either flat or declining oil consumption. Residences and commercial enterprises that survived the 1970s break point became far more efficient and frugal. Factories posted notable gains too. However, the slowing down in industrial oil demand in recent years has been the result of the United States migrating toward a knowledge-based service economy, rather than changes in national energy policy.

In fact, America's growing oil dependency is overwhelmingly rooted in the automobile. The nation must look there to boost competitiveness across the entire economy.

Asking people to buy smaller vehicles hardly generates much enthusiasm or action from the masses. There is always a socially or environmentally conscious segment of the population that is willing and even eager to make the sacrifice, but this is not enough to make a big difference. Market forces, through higher prices, do help people become more conscious of the problem by lightening their wallets. But in the United States it's still difficult for people to voluntarily trade in their vehicles for ones that get, say, 25 percent or more fuel economy. Putting aside the nontrivial issue of perceived safety and the mentality that I-need-an-SUV-to-haul-all-my-gear-around, the cost of trading up for a new fuel-efficient vehicle is large. A $24,000 vehicle that is four years old has typically depreciated down to half its original value. In the absence of subsidies or a strong social conscience, a driver that is being asked to trade in for a vehicle with higher fuel economy has to recoup $12,000 in gasoline savings in a reasonable period of time to make the switch worthwhile.

At $2.50 per gallon of gasoline, a driver with average habits commuting 12,000 miles per year will save $42.00 per month by trading up to a vehicle that gets 30 miles to the gallon instead of 20. At $4.00 per gallon, the monthly savings are $67.00. While those are notable savings, financially it's not enough to sway someone to lay out an extra $12,000 for a new vehicle.

Clearly, market forces are not going to be enough to effect rapid change. Except for those who are socially conscious, most people are likely to hold on to their cars for financial reasons. They will typically wait until they are tired of their current vehicle (on average seven years), and then consider a vehicle that gets substantially better fuel economy. Government-sponsored incentives for buying fuel-efficient vehicles would help, such as the $3,400 tax credit for buying a hybrid provided in the new U.S energy plan, but the number of people who will likely exercise this option is still too few to yield meaningful change.

Indirectly, reducing road fuel consumption is important for the competitiveness of the nation's industrial base, which is quite sensitive to higher oil prices and price volatility. If drivers hog a large fraction of the oil, the industrial base is forced to pay more. And while 40 or 60 bucks a month in fuel savings doesn't sound like much at the individual level, the national economic benefit of freeing up that much cash is large: 230 million registered vehicles times $40 per month is over $9 billion a month, or $110 billion per year. That's a lot of consumer spending that could grease the economy in productive and globally competitive ways. Of course, there are also other direct benefits in terms of a better environment, or, conversely, the added indirect costs of a dirty environment and volatile climate, if things keep going as they are.

Forcing people to change through legislation is politically risky. Slapping on a hefty fuel tax at the pumps has been a successful policy tool in other countries, but to really work, such a policy requires the provision of alternative modes of public transportation.

It's difficult to think of a more contentious piece of legislation in America than raising fuel taxes. Though building more public transportation is always possible, it's doubly difficult now that there has been a mass migration to the suburbs over the past 20 years.

So what has to happen to curb road fuel demand in America? As discussed in Chapter 6, the options for change are limited and very slow moving if preservation of lifestyle remains a priority, but the possibility opens up considerably when that restriction is taken away. In this sense, softening up the nation's defenses against lifestyle change are key to solving our energy problems.

There are two things that speak louder to people than their wallets: a cry to rally around the flag for the good of the nation and social peer pressure. Rally cries have been instigators of big change in many countries around the world as well as in the United States. In the United States, the most recent rally cry has been about the War on Terrorism. Unfortunately, as with the War on Terrorism, a crisis is usually necessary before a rally cry will be sounded, let alone heard. The current pressure buildup in global oil supply chains may offer governments the necessary support to get things going, but any rally cry will have to be accompanied by legislation that guides the population through the break point to new energy consumption habits. That's why a rally cry is needed. It's up to the government to decide what legislative tools will work best. Beefing up incentives for efficiency and conservation, imposing lower speed limits again, raising fuel taxes, applying a progressive tax on vehicles based on fuel inefficiency, and odd-even license plates schemes are among the many changes that can be imposed once the rally cry is sounded.

The second force, social peer pressure, is possibly stronger, although more difficult to predict or influence. Initiating a movement that would make driving large cars socially unacceptable would be very powerful, but is it possible? In the last 10 years, smoking has become frowned upon, and many large cities have instituted smoking bans in public places like restaurants and bars. This was inconceivable even a few short years ago. So, too, driving large vehicles could be viewed as irresponsible because it is not good for the health of the nation. Grassroots social movements can be more powerful than pure market forces for bringing about large-scale change and often work in advance of changes in the law. In this sense, environmentalist and other altruistic lobby groups would do well to steer their efforts towards public education campaigns about conservation and efficiency rather than into efforts to block the building of new nuclear power plants, refineries, LNG terminals, or pipelines. We need a well-balanced and well-supported energy supply infrastructure in order to have a "healthy" economy, in every sense of the word. We need energy to power our hospitals, light our education facilities, keep our environment clean, and grow our quality of living. Today, and over the next 20 years, the real frontline in the battle for reducing energy dependency, wastefulness, and negative environmental impact is with the individual member of society, even more so than the industrial power producer or corporate consumer.

The problems in the United States are complex and lack easy solutions. Market forces will not be enough to drive rapid change in U.S. consumption of road fuel, the root of America's dependency problem. Legislation is difficult to enact in the absence of a rallying cry, because first there must a compelling event that inspires such a cry, and the cry itself must be accompanied by strong government inducement to change lifestyle habits. Social forces—as nebulous and indirect as they might be—are the most effective tool of all. A combination of social education and social engineering along with good energy policy are the solution to a big problem that is getting progressively worse.

China

Viewing China's growth through Reid Sayers McBeth's eyes, the future looks like a rocket ride straight to global disaster. But it is a mistake to "straight line" China's oil consumption in proportion to its population or its current rate of economic growth, or both. Neither scenario paints a sustainable picture of oil consumption nor is representative of how an industrializing and then maturing nation's long-term oil consumption evolves. China's dual-fuel mix of coal and oil is on a path of diversification. Big hydroelectric projects, up to 40 new nuclear power plants by 2025, and many large LNG and natural gas pipeline projects will start to have a noticeable impact on the country's energy mix by 2012. Industries will become gradually more efficient. As in every other aggressively industrializing economy that has gone down the same path, China's economic growth will gradually become less dependent on oil. How much less depends on policy, which is difficult to predict for any country, never mind China.

The extent and depth to which energy policy is implemented in China will depend on how high and how fast commodity prices rise and how hot the geopolitics for oil become in the current great scramble. The higher oil prices go, and the more tense the global scramble becomes, the sooner we can expect policy actions that signify a break point.

Higher fuel taxes and driving limits such as speed and odd/even license plate restrictions are the more typical measures used by nations to begin to curb road fuel use. But there are other alternatives available to a country like China, where only a relatively small fraction of the population (8 of every 1,000 people) currently own cars. It's daunting to think about 1.2 billion people suddenly buying and driving cars over the next few decades. But high-growth countries like China and India have the benefit of having studied what other countries have done in their growth periods, and they also have the benefit and the promise of modern technologies. In addition, the supply chain infrastructure is not yet fully developed across the country, nor is there a legacy fleet of fuel inefficient vehicles. For these reasons, China in its formative years of industrialization has a golden opportunity to engineer a society that does not fully experience the level of oil addiction that we have known in the West. Admittedly, from today's standpoint it doesn't look like the opportunity is being seized, but then again the break point has not yet been reached.

For the foreseeable future, China is committed to using oil to propel its progression to an economic superpower. When they'll decide to ease back on consumption remains to be seen, but China's application to host the Olympics in 2008 provides a glimpse of what's to come. In its energy plan memo to the Olympic Committee, the Chinese government promised to "honor the commitments to the energy and environmental protection actions made in the bidding report to the IOC." This includes overhauling Beijing's coal-dominated energy mix and instilling a market-based energy supply system. The memo goes on to outline Beijing's plans for environmental quality and sustainable development: the deployment of new technologies that will rely on cleaner energy, the restructuring of industries to offset the growth of energy consumption, efforts to ensure security of supply through market mechanisms and diversification of supply, and a reproportioning of the share of clean and efficient energy in the overall energy mix. If these policies are applied not only in Beijing but in the nation as a whole, then the Beijing Olympics of 2008 may well produce China's energy break point.

ROW—The Rest of the World

While China and the United States are the lightning rods for debate about demand growth, we should not forget that almost half of the world's new oil consumption is originating from the rest of the world. There is a lot of latitude to rebalance in places like Latin America and Eastern Europe, and particularly in places that are not affluent enough to be paying for high-priced oil. Brazil is worth watching as well, because it is on track to become a net oil exporter in the next few years.

Japan, South Korea, and Western Europe will continue to lead in the prudent use of oil in their economies. Observing South Korea 10 years ago, before their currency devaluation, you would have been inclined to straight-line their oil consumption to the moon. Who would have thought that nation could instill diversification of its energy mix such that demand for oil is now almost unlinked to its high-growth economy? South Korea shook its unchecked addiction to oil around 1999. Along similar lines, most experts are amazed to learn that Japan, Britain, France, and many other European countries do not consume any more oil today than they did in 1973—the last break point. By rallying a nation and using effective energy policy and creating a social stigma about wasteful consumption, long-lasting change has been achieved. These are the changes that will see visionary nations of the world through the next break point. Waiting idly for normal market dynamics to effect change will simply not be good enough.

Business

Leaders of corporations that consume a lot of energy have a golden opportunity today to differentiate themselves by becoming low-cost producers within their peer group. While dinosaur competitors wait for energy prices to fall, industry innovators can separate themselves from the Jurassic herd by investing in more energy-efficient processes. And the time to start is now.

Corporations that are energy intensive—companies in industries like steel, forestry, and any manufacturing sector where energy is a substantial fraction of operating costs—have the biggest chance to jump ahead of competitors. The first step for managers that run energy-intense companies is to recognize that world oil prices and North American natural gas prices are not coming down any time soon. Industry groups that I have met with seem to be playing a wait-and-see game. They are holding out to see if oil and natural gas prices are going to drop before deciding to invest capital in processes that will modernize their plants to be more energy efficient. Their reluctance is, in part, good business sense: up-front capital costs can be high, and short-sighted shareholders can be brutally unforgiving to managers that expend capital on technologies that do not provide quick returns. In addition, current market volatility means that energy prices will occasionally drop in the short term, giving laggard corporations permission to continue on as if all is well.

Motivation for proactive change comes with the realization that higher average prices are here to stay. In other words, business leaders must be convinced that today's pressure build is not a short-term phenomenon before they will act. Unfortunately, inertia and a "prove it" mentality will keep many companies from getting ahead of the pack and becoming an industry leader.

In 2004, I gave a speech on energy issues to an industry group. At the end of the talk, a middle manager of an industrial company approached me and asked to talk in more detail about what was going on with oil and natural gas prices. It was budget time and his boss wanted him to put together the numbers for his division. He wasn't an expert in oil and gas and never had to worry about high and volatile prices before. But petroleum products had quickly grown into the largest expense in his budget, and he didn't know what prices to use going forward. All the media and analyst chatter was giving him mixed signals about whether prices were going up, down, or sideways. Many in his company were of the belief that prices were going to come down soon. At the risk of contributing to his confusion, I gave the manager a detailed briefing of why I felt oil prices were not coming back down to $20/B, and natural gas prices were not going back down to $3.00/MMBtu. He thanked me and returned to his office to finish his budget. Did he use the higher numbers that would have come from my projections? I have no idea, but I do know that the pressure to use lower estimates would have been intense. That manager is not alone in his confusion. Many executives whose companies rely heavily on energy are not experts on energy. After 20 years of low prices, they don't know how to plan their corporate strategy to deal with volatility. The easiest thing to do during a period of volatility is to try to wait it out. Today, that is exactly the wrong thing to do.

As a manager of an energy-intense business—whether directly or indirectly affected by higher oil prices—the right thing to do now is to take advantage of the confusion among your peer group competitors and make immediate changes toward becoming the low-cost producer of tomorrow. Instead of complaining about energy costs and poor bottom-line results, managers should start making their industrial processes more energy efficient and therefore more profitable. Any such investments will be capital well spent. In a competitive sense, the current situation is a gift to those astute business leaders who understand that their energy input costs are neither going to stabilize nor fall appreciably. They know that this is a perfect opportunity to differentiate their cost structures from the competition, and that their investments are likely to pay back in an acceptable period of time. More importantly, they can position their companies as energy-conscious producers—the coveted darlings in the market of tomorrow.

Tolko Industries is a company you may not have heard of before, but it provides an example of the kind of industry leadership I'm describing. As a marketer and manufacturer of specialty forest products like wood beams and panels, Tolko consumes a lot of natural gas in its operations. Because of the hit taken to its operating costs, Tolko has been closely following the increase in natural gas prices, which in North America are rising directly in proportion with oil prices. Back in 2004, Tolko realized that oil and natural gas price increases were here to stay and decided to do something about reducing the impact of higher prices on its operating costs before it became too severe. To do so, they sought out the advice of Nexterra, a small entrepreneurial company that makes bio-reactors. Bio-reactors can produce a clean-burning synthetic natural gas from wood chips. It was ideal for Tolko, because as a forestry company it produces bark as a by-product of its operations—something that it obtained little revenue from.

As natural gas costs started rising, management recognized the much higher value of the bark as an "alternative" fuel. Given the built-in bark supply, buying a bio-mass reactor from Nexterra made a lot of sense. Any manager could have made such a decision, right? Not exactly. Bio-mass reactors cost a few million dollars—too expensive if fossil fuels are cheap. The executives of Tolko had to take a leap of faith to outlay millions in up-front costs to build a future competitive advantage against slower-moving peers. Now, the economics of the decision look great: the company is going to save at least $1.5 million per year by rebalancing to a new supply chain fed by renewable bark. That sort of decision making takes foresight and understanding of the world's energy dynamics, and I believe that it has secured Tolko a position as the low-cost producer of tomorrow.

I don't profess to be an expert on forestry, steel, airlines, or any other industry group that is heavily reliant on energy for day-to-day operations. But from discussions that I've had with people in these businesses, my expertise tells me that there is plenty of latitude for companies to creatively save energy and make their operations more profitable for the long term. But the "long term" is the key qualifier. Energy-dependent companies need to look toward the future in order to be prepared for the rebalancing period we are about to enter into.

Entrepreneurs

While most people are resistant to and even intimidated by change, entrepreneurs revel in it. The greater and more radical the changes taking place, the bigger the sandbox that an entrepreneur gets to play in. Why is the entrepreneur having such a good time when everyone around is looking stressed? The entrepreneur knows that all of the confusion and uncertainty is creating big opportunities for making a fortune.

The world's biggest fortunes have been made during times of radical change. The digital revolution has been responsible for a great deal of change and entrepreneurship since the 1980s. Consider the fortunes made by the founders, financiers, and even the early employees of companies like Silicon Graphics, Intel, Microsoft, Yahoo, eBay, Cisco Systems, and Google. Computing, telecommunications, medical imaging, graphic design, software, photography: the number of goods and services that have been profitable in the digital revolution has been unbelievable. We may never see the likes of it again.

Except that we're actually seeing a new revolution germinating right before our eyes in the energy industry. Historically, when we think of the fortunes made in the energy industry, we think of unassailable titans like John D. Rockefeller and J. Paul Getty or behemoth energy companies like ExxonMobil, Shell, General

Electric, and Siemens. These leaders and entities made their fortunes during the break points of the past, as the world cowered around them. While economies of scale remain important in energy, don't be misled into believing that only titans and giants will thrive in the world of tomorrow. In fact, as with the digital revolution, many fortunes will be made by the smaller players who recognize early on the opportunities of a rapidly evolving new game. Opportunity abounds through each phase of the energy cycle: Growth and Dependency, Pressure, Break Point, and Rebalancing. The most radical changes (and greatest opportunities) present themselves when society transitions from the fog of the pressure buildup, through the intimidation of a break point, and into the clarity of rebalancing and renewed growth. The least amount of change and opportunity is found when growth is slowing, supplies of incumbent fuels are plentiful, and pressure is low.

We're still in the pressure build phase, so one area where entrepreneurship is buoyant is what is known as the junior oil and gas sector. Like high-tech treasure hunters, these so-called "junior independents" are small, upstart companies composed of a handful of people that have the technical expertise to explore for and develop oil and gas reserves. Most of those people have left larger independent oil companies and gone off on their own. The reasons behind this migration of talent and capital are quite simple. Oil reserves all over the world are maturing, so the pool sizes of those reserves are becoming smaller. The large elephant fields are becoming increasingly scarce and remote. The super-majors like ExxonMobil, Chevron, and Shell are abandoning mature fields and chasing whatever elephants remain because they need to offset their massive production declines and grow their output to satisfy their shareholders. This trend has been particularly strong in the United States and Canada because we've been exploiting our reserves for 145 years, but it's also happening in places like the North Sea.

The fact that oil reservoirs in most parts of the world have become too small for a super-major company to chase does not mean that those pools are no longer exploitable. In fact, for a company of the right size, they can be quite lucrative. Over 1,000 junior oil and gas companies are now headquartered in Calgary, Alberta, where a great deal of the world's technical expertise in the oil and natural gas industry is concentrated. My firm, ARC Financial, has a mandate to invest in entrepreneurs anywhere on the energy landscape. Our principal domain is investing in early-stage companies that explore for oil and natural gas, as well as those companies that build energy infrastructure or exploit non-conventional sources of oil and gas. Business opportunities abound to say the least. The fact that the opportunities for investors are so positive now is symptomatic of the strain and pressure building up in the evolutionary energy cycle.

The rapid growth of the junior oil and gas sector is an early trend in the wave of opportunities still to come. Fortunes will be made picking the carcass of the old energy industry and tapping into the technology and services that will typify the emerging energy industry. Just because fuel cells or ethanol won't alleviate the intensity of our demand for light sweet crude on a national or global scale does not mean that it won't be lucrative building or supplying a better mousetrap to a niche market. Ask Nexterra, a company that is meeting such needs, one mousetrap at a time.

Investors

The first question that investors ask me is, "How will I know when the peak of the oil market has been reached?" The answer is not straightforward and requires some qualification. What is the peak? Are we talking about peak oil prices? Peak oil production? Peak of the cycle? What cycle? The break point?

Usually, it's price that everyone is interested in. I can tell you that strong oil prices will persist as long as the market is expecting aggressive year-over-year demand growth in the face of challenging supply. We know that an economic slowdown will bring down oil demand expectations pretty rapidly. A drop in world GDP growth to 2.5 percent from 4.4 percent would shave about 700,000 barrels per day off incremental year-over-year demand in today's environment. Assuming oil production dynamics don't change, a major economic slowdown would be enough to bring oil prices down pretty hard, quite likely under $40.00 per barrel again. So, an obvious cue to knowing when the oil market has peaked is to follow the broad economy and understand when it is going to slow. But that's neither new, nor illuminating advice.

For oil prices to have really peaked from a long-term perspective, the current cycle has to go from its pressurized condition to a break point, and then onto rebalancing. Only then will the dynamics of oil have truly changed. Merely easing pressure through economic slowdown is an insufficient condition for concluding that oil prices have peaked and dependency problems have been solved. From a societal perspective, the worst thing that could happen right now is an economic slowdown that precedes the break point. Most people will breathe a sigh of relief and think that everything that has happened over the past three years was a false alarm or some sort of conspiracy by the oil companies. In fact, a premature economic slowdown can be better characterized as a "false break point." Note that a false break point, leading to a lower price for oil, will remove needed financial incentive for investment on the supply side. Exploration programs, infrastructure projects, and investment in the large-scale classical alternatives like coal would likely slow. Pressure will ultimately build again, as soon as the economy revs up once more, or as soon as oil companies start reducing their capital investments, or both, for a double whammy.

But investing in energy is not all about watching oil prices flickering on a computer screen, trying to figure out the precise peak, or forecasting next year's average price to the nearest dollar— these are things we know that no investment analyst or economist can do with any consistency. Further, price is merely one gauge, measuring one facet of pressure, in an incredibly complex supply chain. It may be telling you that the engine boiler is about to blow off steam, but the really valuable information is in knowing where the train is going to end up.

Successfully investing in energy, or any other industry for that matter, is about anticipating the type and character of changes to come. Changes can be short or long term. Energy investing offers the spectrum: from the minute-by-minute trading of oil futures to the decades-long development of the next greatest thing since the lightbulb. Whatever the case, I purposely wrote this book with a "top down" philosophy that sought to give both long- and short-term investors not specific advice about what the price of oil is going to be in each of the next 10 years, but rather instilling a way of thinking about the dynamics of energy, and the work we derive from it. If you understand that, you understand investing opportunity.

So I come back to my Evolutionary Cycle of Energy, the one I discussed back in Chapter 1. As the dynamics of energy evolve around the cycle there is plenty of opportunity. Here is some broad advice on what to focus on in each of the two halves, as depicted in Figure 7.1.

Growth and Dependency, Pressure Buildup

Value is created in the right half of the evolutionary cycle by finding and developing the energy resource, bringing it to market, and putting it to work.

Resource assets—When growth is ramping up, society has become addicted, and pressure is building, then primary energy resource assets are increasing in value. This is McBeth style investing. In other words, companies that own oil assets appreciate in value if the oil supply chain is tightening up. The same applies for coal, natural gas, or uranium. Often, because primary fuels can substitute for one another—for example, both fuel oil and natural gas can be burned to generate electricity—the tightening up of one commodity adds value to the substitute in anticipation of substitution and rebalancing. Investing in low-cost producers is key; high-cost enterprises are most vulnerable to value loss after the break point.

Resource services—Somebody has to get the resource out of the ground, and in the oil and gas business it's not usually the same companies that own the reservoirs. Companies that own and operate drilling rigs, compressors, pipelines, transport trucks, tankers,

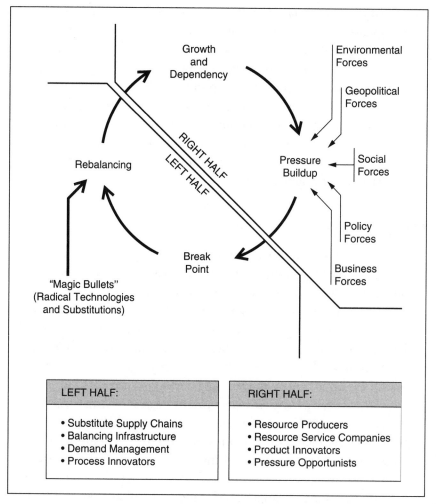

Figure 7.1   Energy Evolution Cycle

and the myriad of other services that support the thirsty supply chain become very active as the scramble to bring on more supply heightens.

Specialty manufacturers—The energy business is a vast mesh of steel, equipment, and hardware. Such products are in high demand to bolster the supply chain infrastructure from beginning to end when there is an urgency to bring more and more fuels to market.

Product innovation—Finally, during the growth and dependency phase there is opportunity to invest in companies that are pioneering innovative new products that convert energy. When electricity was brought to market it spawned a mind-boggling array of new products ranging from lightbulbs to hair dryers, and companies are still innovating today. Gasoline's versatility triggered much more than the automobile. Leaf blowers, chain saws, and campsite generators are a small subset of product innovations that have offered entrepreneurial companies and their investors the opportunity of financial gain.

Break Point and Rebalancing

After the break point the emphasis changes. Consumption growth for the disadvantaged fuel levels off. Nations, companies, and individuals try to rid themselves of the dependency. The value of the energy resource asset peaks and starts declining. The emphasis shifts to rebalancing solutions and other substitute resources. Things are still changing and there is still money to be made. But invest in rock oil, not whale oil. Buy drilling rigs, not whale ships. Search for opportunity in the way people conserve.

Substitute resources and supply chains—Liquefied natural gas, coal, uranium, and renewable energy sources will all gather value as light sweet crude oil becomes progressively encumbered by pressure and ultimate break point. Nonconventional sources of petroleum products like oil sands and shales will also remain at the fore. This often happens well before the disadvantaged fuel, in this case oil, hits its break point. So playing this investment space does require foresight and monitoring of many "pressure gauges." Remember, it's not necessary to play the evolutionary cycle sequentially in time; what's important is understanding the dynamics, anticipating the changes, and predetermining the most likely outcomes. Look for the next best substitute to the disadvantaged fuel, because it will gain most in relative value. Right now, since there are no "magic bullets," coal, natural gas, and uranium all look like high-value winners. Don't forget renewable supply chains too.

Balancing infrastructure and services—Building infrastructure to facilitate bringing new primary fuels to market requires massive investment in infrastructure. Investing activity in liquefied natural gas infrastructure and Canadian oil sands highlights this point. Look for companies that are building equipment, hardware, and specialty devices that facilitate rebalancing toward alternative energy supply chains.

Demand management—When fuel prices go up, consumers complain. Then they start minding how much they consume. Keeping tabs on conservation and counting dollars becomes important. The digital revolution is spawning the means to do so. Information-gathering devices to monitor energy consumption, energy efficiency, and dollar consumption are going to become increasingly important in all sorts of day-to-day appliances. From corporations to individuals, we'll all be much more aware of how much the work we extract from our various fuel sources is costing us. Companies seizing this opportunity will expand the digital revolution to fertile ground still steeped in the analog era.

Process innovation—After decades of product innovations in the growth phase of a fuel have made entire nations addicted on novel ways of extracting work, something must be done to alleviate the pressure without sacrificing the amount of work. We know that rebalancing is not just about switching to alternatives and conserving. It's also about finding process innovations that improve the efficiency of all the hitherto developed products extracting work out of our primary fuel supply chains. Technology is a major enabler of energy efficiency, and companies that are pioneering ways to do so have an increasing value proposition as pressure builds and a break point is reached. Further, the energy policies of many countries, including the United States, are offering grants and subsidies amounting to billions of dollars for companies to find process solutions. Product innovation is giving way to process innovation, and investing opportunities will be increasing in this area.

Magic Bullets

Investing in radical new technologies is a high-risk, long-term proposition. Even more so today than in the past, because we've used up the obvious magic bullets and we're pushing the theoretical efficiency limits on many mainstream processes. Anyone, or any company, purporting to have a magic bullet today that solves all of our energy problems (as rock oil did for whale oil) should be viewed with a high degree of skepticism. By now you should have a sense of the massive capital investment and long time horizons involved in establishing even a toehold on an energy landscape entrenched with long-standing standards and legacy hardware. The odds of winning on any such investment have gone down drastically over the last 150 years.

Individuals

And what of the individual, the small business owner, office worker, doctor, lawyer, business executive, artist, nonprofit director, factory worker, teacher, parent, student, or retiree? All of the guidance provided in the above sections affects their world to varying degrees. Perhaps it is part of the dynamics that will shape their lifestyles or workplaces; perhaps it will change the tax and regulatory regime or business climate that they occupy; perhaps the investment opportunities will influence their thinking and allow them to understand the opportunities and dangers inherent in the changes ahead. Just as many household investors wish they could have better predicted the ups and downs of the NASDAQ, the energy economy promises to be a roller coaster ride too.

At every energy crisis, it seems that individuals receive the bulk of the pain, feel the burdens most keenly, and get the same well-worn recommendations for weathering the storm. As gas and heating prices rise, we'll all feel the hit to our wallets. The solutions that have worked in the past will work again in the near future: invest in ways and habits to conserve your energy use; buy smaller, more fuel efficient vehicles; upgrade to more energy-efficient appliances. You've heard those tips before, and you'll be hearing them more and more in the future. My advice is to take heed, the sooner the better. If you're buying a vehicle in a few months, forego the gas-guzzling SUV and buy the fuel-efficient sedan. If you're buying a house, check out the furnace and the energy costs closely before making the decision. If you're choosing between whether to build a deck or insulate the attic, realize that energy prices are not coming down.

I can't predict the rocky and smooth waters that your life will travel in the next 10 to 20 years, any more than I could predict that a farm boy in New England in the mid-1800s would end up signing on as a whaleman in Nantucket, heaving barrels of crude oil in Pennsylvania, or replacing the candles in his textile factory with kerosene lamps while making room next to the waterwheel for a new, coal-fired steam engine. But I can predict that knowledge and understanding about the confusing times ahead will benefit your life in myriad ways and influence the choices that you make. Our expectations for the future, whether the economy will become robust or teeter along, whether inflation will remain low or start to rise, whether our nation or industry is in a good position or a bad position, depend more on energy prices than we realize. Those of us who save money through conservation or efficiency, take advantage of new approaches, and weather the storm will be helping ourselves and our families. At the same time, we will be helping our country become more secure, economically stable, environmentally sound, and competitive—a healthier, more prosperous, more opportunity-rich place to be.

The Way Forward

As I write these final words, it's nearing the end of the summer of 2005. Oil has topped $65 a barrel, and the winter season is just around the corner. The aftereffects of hurricanes Rita and Katrina has are still disrupting refinery operations and distribution in the

Gulf of Mexico, reminding a jittery market, yet again, how vulnerable we are, operating with so little spare capacity. Politicians in the United States, newspaper editorials, and some business leaders are still decrying China National Oil Company's bid for Unocal because of the national security issues at stake; meanwhile CNOOC has calmly upped its bid. I've just come back from a two-week family vacation in Europe. A few years before, we traveled to the Eastern seaboard , where we took a detour to visit some water-powered textile mills. The summer before, we traveled to England, where we saw the birthplace of the Industrial Revolution. This year, we had no energy agenda in mind, and yet energy issues would be as impossible to ignore as ever.

The first day we were in London, terrorists exploded four bombs at coordinated locations throughout the public transportation system, killing 52 innocent people and wounding hundreds more. London, hit by the shrapnel of geopolitical tensions many times in its history before, wept, mourned, and very quickly got back to the business of everyday life. Even the financial markets, sensitive to, but weary of the frequent volatility of our modern times, only slowed down for a day and barely registered any alarm. We're getting used to sudden change and the unexpected, after 20 years of spectacular, optimistic growth.

My family and I wandered around London the day of the bombing, avoiding public transportation and making our way on foot to the various tourist sites as police and ambulance sirens wailed incessantly in the background. As we observed the resiliency and calm of the British, I recalled the foresight of Winston Churchill, who had shifted his nation's economy, and thus the world, from coal to crude oil. Before the day was out, we found ourselves at Westminster Abbey, where many heralded figures in science, politics, and the arts like Isaac Newton, Benjamin Disraeli, and Charles Dickens have been buried or commemorated. Marveling at the names, I found myself standing before a commemoration of James Watt. Reading the inscription, I reached for my pen and copied it down: "The King, his ministers, and many of the nobles and commoners of the Realm raised this monument to JAMES

WATT who directing the force of an original genius, early exercised in philosophic research, to the improvement of the steam engine, enlarged the resources of his country, increased the power of man, and rose to an eminent place among the most illustrious followers of science and the real benefactors of the World."

Enlarged the resources of his country; increased the power of man. Such simple words, yet what a world emerged from the energy that Watt showed us how to harness. So too with all of the innovators and leaders who have helped us find, exploit, and benefit from the energy resources of our planet. We are fortunate when we are able to take that energy for granted because it means that we live in stable and prosperous times. Those days are over for the moment. High and volatile prices are now the norm in this chapter of our energy evolution. But we shouldn't forget that the turmoil and uncertainty surrounding energy throughout history has always led to a brighter future. I believe that pattern will hold true in the era we are entering now.

Notes

1   *New York Times,* July 12, 2005.

BIBLIOGRAPHY

Anderson, J. W., *Diesel Engineering*, New York, McGraw-Hill, 1933.

Asian Energy Markets: Dynamics and Trends, The Emirates Center for Strategic Studies and Research, Abu Dhabi, 2004.

Baker, Robert L., *Oil Blood and Sand*, New York, D. Appleton-Century Company Incorporated, 1942.

Barlow, Raymond E. and Kaiser, Joan, *A Guide to Sandwich Glass Whale Oil Lamps and Accessories*, Wyndham, NH, Barlow-Kaiser Publishing Company, Inc., 1989.

Billington, David P., *The Innovators: The Engineering Pioneers Who Made America Modern*, New York, John Wiley & Sons, Inc., 1996.

Boyle, Godfrey. *Renewable Energy; Power for a Sustainable Future, Second Edition*. New York, Oxford University Press in Association with The Open University, 2004.

Brannt, William T. *The Manufacture of Soap & Candles*. London, Sampson Low & Co., 1888.

Bright, Arthur A. Jr. *The Electric-Lamp Industry: Technological Change and Economic Development from 1800 to 1947*, New York, The MacMillan Company, 1949.

Brunner, Christopher T., *The Problem of Oil*, London, Ernest Benn Limited, 1930.

Chesterman, John I., *An Index of Possibilities: Energy & Power*, New York, Pantheon Books, 1974.

Comfort, Darlene J. *The Abasand Fiasco: The rise and fall of a brave pioneer Oil Sands extraction plant.* Jubilee Committee, Fort McMurray, 1980.

Conant, Melvin A. *The Universe of Oil: Selections from the Geopolitical Writings of Melvin A. Conant.* Calgary, Canadian Energy Research Institute, 1999.

Creighton, Margaret S. *Rites & Passage: The Experience of American Whaling, 1830-1870.* Cambridge UK, Cambridge University Press, 1995.

Danielsen, Albert L. *The Evolution of OPEC.* Harcourt Brace Jovanovich, Inc., New York, 1982.

Deffeyes, Kenneth S. *Hubbert's Peak; The Impending World Oil Shortage.* Princeton, NJ, Princeton University Press, 2001.

de Mille, George. *Oil in Canada West, The Early Years,* Calgary, 1969.

Denny, Ludwell. *We Fight for Oil.* New York, Alfred A. Knopf, 1928.

Dickens, Charles. *Bleak House.* Oxford UK, Oxford University Press, 1991.

Dunn, Seth. *Hydrogen Futures: Toward a Sustainable Energy System,* Washington, D.C., Worldwatch Institute, 2001.

Egloff, Gustav. *Earth Oil,* Baltimore, The Williams & Wilkins Company, 1933.

Eveleigh, David, J. *Candle Lighting,* Princes Risborough, Buckinghamshire, Shire Publications Lts., 2003.

Fanning, Leonard M. *Foreign Oil and the Free World.* New York, McGraw-Hill, 1954.

Fischer, Louis. *Oil Imperialism.* New York, International Publishers, 1926.

Forbes, R.J. *Studies in Early Petroleum History.* Leiden, Netherlands, E.J. Brill, 1958.

Friedel, Robert, and Israel, P., *Edison's Electric Light: Biography of an Invention,* New Brunswick, New Jersey, Rutgers University Press, 1986.

Golley, John, *Genesis of the Jet: Frank Whittle and the Invention of the Jet Engine,* Shrewsbury, England, Airlife Publishing, 1996.

Gould, Ed., *Oil; The History of Canada's Oil & Gas Industry.* Hancock House Publishers, Surrey, 1976.

Grayson, L.E. *National Oil Companies.* New York, John Wiley & Sons Ltd., 1981.

Hawken, Paul, Lovins A. and Lovins L., *Natural Capitalism: Creating the Next Industrial Revolution,* Boston, Little Brown and Company, 1999.

Hemsley Longrigg, Stephen. *Oil in the Middle East: Its Discovery and Development.* Issued under the auspices of the Royal Institute of International Affairs, Oxford University Press, London, 1954.

Hough, Walter. *Collection of Heating and Lighting Utensils in the United States National Museum, Bulletin 141*, United States Government Printing Office, Washington, 1981.

International Energy Agency. *Energy Policies of IEA Countries—The Republic of Korea 2002 Review*, IEA Publications, Washington, 2002.

Israel, Paul., *Edison: A Life of Invention*, New York, John Wiley & Sons, 1998.

James, Peter, and Thorpe, N., *Ancient Inventions*, New York, Ballantine Books, 1994.

Kennedy, William J., *Secret History of the Oil Companies in the Middle East, Volumes I and II*, Salisbury, N.C., Documentary Publications,1979.

Kugler, Richard C., *The Whale Oil Trade 1750-1775*, New Bedford, The Colonial Society of Massachusetts, 1980.

Leavitt, John F. *The Charles W. Morgan*. Mystic, Connecticut–Mystic Seaport Museum, Incorporated, 1998.

Lumley, John L., *Engines: An Introduction*, Cambridge, U.K., Cambridge University Press, 1999.

Marre, Louis A. *Diesel Locomotives: The First 50 Years–A guide to diesels built before 1972; Railraod Reference Series No. 10*. Waukesha, WI, Kalmbach Publishing Co., 1995.

McLaurin, John, J., *Sketches in Crude Oil*, Harrisburg, PA, Published by the Author, 1896.

Millard, Andre, *Edison and the Business of Innovation*, The John Hopkins University Press, Baltimore, 1990.

Mrantz, Maxine. *Hawaii's Whaling Days*. Aloha Publishing, Honolulu,1976.

Northrup, John, D., *Natural Gas in 1915*, Washington, D.C., Government Printing Office, 1916.

Nye, David E., *Electrifying America: Social Meanings of a New Technology*, Cambridge, MIT Press, 1990.

O'Connor, Harvey. *The Empire of Oil*. New York, Monthly Review Press, 1955.

Pope, Franklin Leonard. *Evolution of the Electric Incandescent Lamp*. New York, Boschen & Wefer, 1894.

Ratcliffe, Samantha. *Horse Transport in London*. Tempus Publishing Limited, Stroud, 2005.

The Rushlight Club. *Early Lights, A Pictorial Guide*. The Rushlight Club, 1979.

Sampson, Anthony. *The Seven Sisters: The Great Oil Companies & The World They Shaped.* New York, The Viking Press, Inc., 1975.

Sayers McBeth, Reid. *Oil, The New Monarch of Motion.* New York, Markets Publishing Corp., 1919.

Shwardran, Benjamin. *The Middle East, Oil and the Great Powers.* New York, Frederick A. Praeger, 1955.

Simmons, Matt. *The Oil World: 1973 Compared to 2000.* From Web site.

Soloman, Brian. *The American Diesel Locomotive.* Osceola, WI, MBI Publishing Company, 2000.

Stivers, William. *Supremacy and Oil—Iraq, Turkey and the Anglo-American World Order, 1918-1930.* New York, Cornell University Press, 1982.

Talbot, Frederick A. *The Oil Conquest of the World.* Philadelphia, J.B. Lippincott Company, 1914.

Utterback, James, M., *Mastering the Dynamics of Innovation: How Companies Can Seize Opportunities in the Face of Technological Change,* Boston, Harvard Business School Press, 1994.

Yergin, Daniel. *The Prize: The Epic Quest for Oil, Money and Power.* New York, Simon & Schuster, 1991.

Web Sites

http://www.nps.gov/lowe/loweweb/Lowell_History/prologue.htm

INDEX

Mesopotamia, 39, 41, 47. *See also* Iraq
methane, 171, 200
Mexico, 40, 42, 45, 129
Microsoft, 165
Middle East, ix–xi, 40, 47, 68, 148, 186
  Arab nationalism and, 71
  As-Is Agreement of 1928 and, 69
  demand, consumption, and, 134–135
  OPEC creation and, 70, 73
  "open door" policy for, US
    proposal of, 42, 48–50
  Red Line Agreement and, 48–50,
    49, 69–70
  San Remo Agreement and, 42,
    47–48
  Sykes-Picot agreement and, 41–43
Middlesex Canal, Massachusetts, 152
Mobil, 48, 54, 69. *See also* Socony
  Vacuum
Moby Dick, 13. *See also* whale oil,
  history of, 13
Monthly Oil Report, 143
Morgan, Charles W., 14, 17, 21
Morgan, J.P., 161, 163
Morocco, 36
Morse, Samuel, 157
Mossadegh, Mohammed, 70–71
Mosul, 40–42
Mr. Five-Percent. *See* Gulbenkian,
  Calouste, 48
multinational oil companies, 68–76,
  94, 176
  Bolshevik revolution, Russia, 70
  embargo of 1970s and, 75
  independent oil companies vs., 71–72
  Mossadegh administratio of Iran
    and, 70–71
  OPEC and, 73–74
  Suez Canal Crisis of 1956 and, 71
multipolar world view, xi, xiv

National Maximum Speed Limit Act
  of 1974, 90, 184, 192
national oil companies (NOCs), 138–139
natural disasters, impact of
  Hurricanes Katrina and Rita
    (2005), x, 224, 253–254
natural gas, xii–xiii, 1–2, 8, 23, 86–87,
  104, 167, 168, 170, 171, 175,
  182, 198–200, 228–229, 239,
  246, 250–251

Nature, 155
Nazi Germany, 22
New England and whale oil, 8–19
New York Herald Tribune, 94–95
New York Times, 163
Newcomen, Thomas, 29, 63, 233
Newcomen Steam Engine, 29, 30
Newfoundland, 120, 141
Newton, Isaac, 254
Nexterra, 243, 246
Nigeria, 84, 135, 136, 211
NIMBY lobbies, 174, 212, 231
Nixon, Richard, 59, 76
Nixon Administration, 74
Nobel brothers, 43, 46
North Sea Brent grade, 3, 96–97
North Sea oil fields, 86, 121, 186, 245
Norway, 129
nuclear energy, 20, 21, 23
nuclear fusion, cold fusion, 153–156,
  170
nuclear power, xiii–xiv, 79, 81, 86–87,
  89, 104, 129, 148, 153–156, 167,
  168, 170, 182, 196–197,
  206–207, 227–229
nuclear weapons proliferation, xi–xii

Occidental Oil Company, 73
Oil and Natural Gas Corporation
  (OGNC), 138, 139
Oil Creek, Pennsyvlania, 18, 19
"oil in place," 124
oil sands, 127, 201–204, **202,** 209,
  226, 250
oil shales, 127, 204, 250
oil well, **122**
oil, historical use of, 35–37
Oil: The New Monarch of Motion, 60–61
Oklahoma, 2, 43
Ontario, 18–19
OPEC, 4, 5, 12, 22, 68, 70, 77, 94,
  102, 154, 226, 234
  demand, consumption, and, 139–141
  embargo of 1970s, x, 74–76, 81–90,
    104, 184, 187
  formation of, 73
  multinational oil companies and,
    73–74
  price and, 139–141
  spare capacity and, 129
  Vienna 1973 meeting of, 74

ABOUT THE AUTHOR

Peter Tertzakian is chief energy economist of ARC Financial
Corporation, one of the world's leading energy investment
firms.